Management Extra

# MANAGING HEALTH, SAFETY AND WORKING ENVIRONMENT

AMSTERDAM • BOSTON • HEIDELBERG • LONDON • NEW YORK • OXFORD • PARIS • SAN DIEGO • SAN FRANCISCO • SINGAPORE • SYDNEY • TOKYO

Pergamon Flexible Learning is an imprint of Elsevier
Linacre House, Jordan Hill, Oxford OX2 8DP, UK
30 Corporate Drive, Suite 400, Burlington, MA 01803, USA

First published 2006
Revised edition 2009

© 2009 Wordwide Learning Limited adapted by Elearn Limited
Published by Elsevier Ltd. All rights reserved.

No part of this publication may be reproduced, stored in a retrieval system or transmitted in any form or by any means electronic, mechanical, photocopying, recording or otherwise without the prior written permission of the publisher

Permissions may be sought directly from Elsevier's Science & Technology Rights Department in Oxford, UK: phone (+44) (0) 1865 843830; fax (+44) (0) 1865 853333; email: permissions@elsevier.com. Alternatively visit the Science and Technology Books website at www.elsevierdirect.com/rights for further information

Notice
No responsibility is assumed by the publisher for any injury and/or damage to persons or property as a matter of products liability, negligence or otherwise, or from any use or operation of any methods, products, instructions or ideas contained in the material herein.

**British Library Cataloguing in Publication Data**
A catalogue record for this book is available from the British Library

**Library of Congress Cataloging-in-Publication Data**
A catalog record for this book is available from the Library of Congress

ISBN 978-0-08-055740-3

For information on all Elsevier Butterworth-Heinemann publications visit our website at www.elsevierdirect.com

Printed and bound in Hungary

Working together to grow libraries in developing countries

www.elsevier.com | www.bookaid.org | www.sabre.org

ELSEVIER BOOK AID International Sabre Foundation

# Contents

## Activities

## Figures

## Tables

# Series preface

Whether you are a tutor/trainer or studying management development to further your career, Management Extra provides an exciting and flexible resource helping you to achieve your goals. The series is completely new and up-to-date, and has been written to harmonise with the 2004 national occupational standards in management and leadership. It has also been mapped to management qualifications, including the Institute of Leadership & Management's middle and senior management qualifications at Levels 5 and 7 respectively on the revised national framework.

For learners, coping with all the pressures of today's world, Management Extra offers you the flexibility to study at your own pace to fit around your professional and other commitments. Suddenly, you don't need a PC or to attend classes at a specific time – choose when and where to study to suit yourself! And, you will always have the complete workbook as a quick reference just when you need it.

For tutors/trainers, Management Extra provides an invaluable guide to what needs to be covered, and in what depth. It also allows learners who miss occasional sessions to 'catch up' by dipping into the series.

This series provides unrivalled support for all those involved in management development at middle and senior levels.

## Reviews of Management Extra

*I have utilised the Management Extra series for a number of Institute of Leadership and Management (ILM) Diploma in Management programmes. The series provides course tutors with the flexibility to run programmes in a variety of formats, from fully facilitated, using a choice of the titles as supporting information, to a tutorial based programme, where the complete series is provided for home study. These options also give course participants the flexibility to study in a manner which suits their personal circumstances. The content is interesting, thought provoking and up-to-date, and, as such, I would highly recommend the use of this series to suit a variety of individual and business needs.*

**Martin Davies** BSc(Hons) MEd CEngMIMechE MCIPD FITOL FInstLM
Senior Lecturer, University of Wolverhampton Business School

*At last, the complete set of books that make it all so clear and easy to follow for tutor and student. A must for all those taking middle/senior management training seriously.*

**Michael Crothers**, ILM National Manager

# Creating a safe, healthy and productive working environment

A safe, healthy working environment is good for business. It is not only essential for the well-being of employees, but also for ensuring that organisations are successful and sustainable, and that economies thrive in the long term. Bold statements, but in the course of this book you will examine why a safe, healthy working environment brings economic and personal benefits.

The biggest benefits arise when health and safety is embedded in the culture, reflecting an organisation which values and respects its employees, equipment, materials and products.

There are also financial benefits. The Health and Safety Executive (HSE) has identified companies on its website (www.hse.gov.uk) where specific financial advantages were attributable to improved health and safety:

- One company saved £12 for every £1 it spent
- A 73% reduction in employee insurance claims and 18% fewer days lost to injuries
- A 50% reduction in civil claims

The HSE also calculate the number of days lost to work-related illness every year, and for 2004–5, they estimate:

28 million working days were lost due to work-related ill-health and seven million due to workplace injury.

Source: www.hse.gov.uk

Managers are at the core of the process to develop and maintain an organisational culture that promotes health and safety. We explore the roles and responsibilities that come with a management role and consider safety alongside efficiency. We look at how workplace facilities are managed and how materials and equipment are used, stored and maintained for optimum effectiveness.

Your objectives are to:

- contribute to the culture for health and safety in your organisation
- apply key legal requirements to your workplace
- understand the requirements of health and safety policy, employee consultation and risk assessment for your organisation
- identify how facilities management can contribute to a safe, and effective working environment
- understand the importance of managing equipment and materials purchase, use and maintenance in the workplace.

# 1 Health and safety essentials

The basis of UK health and safety law is the Health and Safety at Work etc Act 1974 (HSW Act). The Act sets out the general duties which employers have towards employees and members of the public. It also covers the responsibilities we have to ourselves and to others. This is a fundamental piece of legislation that governs our responsibilities at work. This theme will look at how the legislation has developed to adapt to new ways of working and to improve the welfare of employees.

The Legislation is founded in common sense, incorporating respect for others and practices which make businesses more profitable and organisations more efficient. It is important to think carefully about hazards, risks and possible improvements and work with colleagues to help them get into the same mindset and create a culture of health and safety.

In this theme you will:

- **explore what health and safety covers and its relevance in your working environment**
- **identify explicit reasons why you need to bother with it**
- **determine the laws you need to be aware of**
- **identify your responsibilities as a manager.**

## What does health and safety cover?

Health and safety is both a management practice and a way of benefiting people at work. It demands attention because it contributes to the production of quality goods and services and a culture of respect for others. On the other hand failure to attend can seriously damage people's health, the organisation's image and you may be breaking the law.

Health and safety is a wide ranging and dynamic area of practice. It may surprise you how much of the day to day life of an organisation is bound by health and safety law and guidance. To illustrate how important it is and to assess the relevance to your area of work, examine the list in the table below, and indicate any areas with a tick, which fall into your responsibility or business remit. There may be others you can think of which are specific to your workplace.

## What it covers

- ☐ Asbestos
- ☐ Asthma
- ☐ Back pain
- ☐ Carriage of dangerous goods
- ☐ Chemical Hazard Information Packaging for Supply Regulations 2002 CHIP3
- ☐ Control of Major Accident Hazards 1999 and Regulations 2005
- ☐ Compressed air
- ☐ Confined spaces
- ☐ Control of Substances Hazardous to Health COSHH
- ☐ Domestic gas
- ☐ Drugs and alcohol
- ☐ Electrical safety
- ☐ Equipment at work
- ☐ Ergonomics
- ☐ Falls from height
- ☐ Fire and explosion
- ☐ First aid at work
- ☐ Genetically Modified Organisms (contained use) GMOs
- ☐ Infections
- ☐ Landuse planning
- ☐ Latex allergies
- ☐ Metal working fluids
- ☐ Moving goods
- ☐ Musculo-skeletal disorders
- ☐ New and expectant mothers
- ☐ Noise
- ☐ Notification of New Substances Regulations 1993 NONS
- ☐ Radiation
- ☐ Road safety
- ☐ School trips
- ☐ Sickness and absence
- ☐ Skin at work
- ☐ Slips and trips
- ☐ Smoking
- ☐ Stress
- ☐ Temperature in the working environment
- ☐ Transport
- ☐ Vibration
- ☐ Violence
- ☐ Working with display screen equipment (DSE)

**Table 1.1** *Range of areas covered by health and safety Acts and Regulations* Source: ww.hse.gov.uk

As you can see, it is not all about dangerous substances and working in controlled environments. It is about making life as comfortable as possible for everyone, from ensuring well fitting headsets in a call centre to safe disposal of by-products from a factory. It applies equally in the office and on a farm. It sets out guidelines and rules which must be adhered to whatever the setting. Cleaning bleaches used in offices are as potentially dangerous as chemicals in a factory depending on how and when they are used and stored.

Companies and organisations that fail to take health and safety seriously are not only at risk from the law, reputations can be made and lost through good and bad management practices related to the issues outlined above.

In law an organisation with more than five employees is required to have a written safety policy statement. This will include the key elements of the law and responsibilities relevant to your organisation's activities. It will cover areas such as buildings, people, machinery, transport, emergency procedures and reporting of accidents. It is important that you are familiar with this document, the key areas of risk and your responsibilities. Obtain a copy of this document and any related policies to support your work on this theme.

## Activity 1

## The scope of health and safety in the workplace

### Objectives

This activity will help you to:

- explore your own and your organisation's attention to health and safety
- understand the importance of the health and safety policies and the key features of health and safety relevant to your organisation's activities.

### Task

1 Obtain a copy of your Safety Policy Statement, read it and if you have any queries discuss them with your nominated Safety Officer. Try to familiarise yourself with the contents at this stage.

2 Look back over the areas you indicated above as important or relevant to safety in your organisation. Try to identify any examples of good practice or good guidelines used by your organisation in relation to these areas, such as noise reduction measures, good hygiene facilities, well-designed furniture to reduce incidences of back pain, environmentally friendly waste disposal.

3 Can you think of any examples of difficulties you or your organisation have experienced in any areas of health and safety?

## Feedback

The aim of this activity is to familiarise yourself with the Safety Policies of your organisation and start to consider the relevance to your particular role. It is worth understanding the positive measures as well as the potential risks. Health and safety law imposes important obligations and restrictions on employers and these may be general to all employers or specific to industries or activities.

# A culture for health and safety

Organisational culture is about the deep-set values and beliefs people share in their work organisations. These values and beliefs provide direction on all aspects of working together, which includes their attitude and behaviour towards health and safety. A safety culture is one where the values and beliefs of the organisation include a total commitment to health and safety.

Management of health and safety is not an add-on to management activities. It is an integral part of working life. Looking after people and property cannot be left to chance. Accidents and preventable injuries cost money and reduce productivity, ultimately hitting the bottom line and the organisation's reputation. People, property, productivity and profits amount to mainstream management concerns.

## The benefits of a strong culture

Organisations with good attitudes to health and safety policy and practices are also organisations with good attitudes to working practices in general and a respect for employees. Quality permeates every area of an organisation with a focus on health and safety. It has become clear from organisations involved in major incidents, such as the sinking of the Herald of Free Enterprise, and the Kings Cross and Clapham rail disasters, that basic faults in structure, climate and procedures may predispose an organisation to an accident.

Similarly unhealthy environments, such as the one showcased by the human resources magazine Personnel Today, are indicative of institutional moves that lead to stress and burn out.

The macho culture in the Police Service is forcing officers to push themselves too hard and undermining health and safety, according to the Police Federation of England and Wales.

There is a pressure on officers that the job has to be done 'at all costs', but it should not be at the expense of people's lives, warned Paul Lewis, secretary of the federation's health and safety committee.

Lewis said management and officers should receive training to give them more of an appreciation of how important health and safety is to help them put in place controls and measures that would complement the job.

Source: Personnel Today, 12, 2004

The belief structures associated with a strong culture are evident in an organisation that promotes:

- leadership and management with a visible commitment to health and safety
- long-term strategy with sustained interest and resources to implement measures
- treating health and safety performance as a business objective
- commitment to health as a positive contributor to productivity
- management practices that put health and safety at the core of a manager' role
- a responsiveness to change
- an accident-free workplace
- working conditions that reduce absence caused by ill-health
- health and safety practices that are people-oriented and customer-focused
- standards and clear procedures for working practices
- communication and co-operation at all levels
- good questioning attitudes and organisational learning
- a good health and safety record.

**Figure 1.1** *Characteristics of a positive health and safety culture*

If you can honestly say that these qualities typify your organisation then you are likely to have an excellent health and safety culture.

### Cost benefits

Good health and safety is rarely a net cost to an organisation. Some organisations have turned the benefits they perceive into stark figures on the profit and loss sheet. Two examples are highlighted below:

**South West Water – Business benefits and cost savings**

In 1991/92 there were 136 accidents per 1000 employees. After the implementation of a programme to promote health and safety this had declined to 53 accidents per 1000 employees by 1995/96. South West Water calculated that it had saved £2,546,000 through its accident prevention measures over the period April 1992 to March 1998.

The company also ran two proactive health programmes and analysed the cost savings using the industry wide representative cost of work-related ill-health. The programmes produced the following projected savings:

- The work related upper limb disorder prevention programme was immediately self-financing, providing loss control savings of £88,500 per annum over the next 10 years.
- The hand arm vibration syndrome prevention programme became self-financing within 2 years, and then provided loss control savings of £19,300 per annum over the next 10 years.

Another organisation, the British Polythenes Industries plc (BPI), calculated that a rehabilitation scheme to support individuals with musculo-skeletal disorders and kinetic handling training resulted in savings of £12 for every £1 spent. Here's how they calculated it.

In 2001, the initiative cost around £16,000. This was the cost of around 400 treatments (with an average of 3 treatments per referral).

For every £1 spent on the treatments, BPI believes they have benefited from savings of £12 because of reduced absences. So, as basic costs in 2001 were £16,000, benefits of £192,000 were achieved. This was a saving of £176,000 in one year and has continued to return such large savings each year since.

Source: www.hse.gov.uk/businessbenefits/index.htm

## Culture at work

Many workplaces are multi-dimensional and multi-locational. Most managers would find it hard to be sure, for instance, that all their field-based sales force are fully compliant with using only hands-free telephones in the car, or taking a break from driving at regular intervals. Managers may not realise the physical and mental stresses and strains involved in a new IT support centre, for instance. The importance of culture is highlighted by these types of situation.

Managers need to take responsibility for workspaces, both within and outside their vision and experience. In a strong culture, which promotes health and safety at work, employees will:

- feel valued
- have sufficient time to do the job appropriately
- have a clear idea of the guidelines set out by the organisation for their health and safety.

Cutting corners is not an option. The signals sent out by a manager who flouts the regulations are that they don't matter, and that the manager has little respect for their staff.

The Health and Safety Executive (HSE) gives the following definition of the safety culture of an organisation.

> The safety culture of an organisation is the product of individual and group values, attitudes, perceptions, competencies and patterns of behavour that determine the commitment to, and the style and proficiency of, an organisation's health and safety management.
>
> Organisations with a positive safety culture are characterised by communications founded on mutual trust, by shared perceptions of the importance of safety and by confidence in the efficacy of preventive measures.

Source: Successful health and safety management HSG 65

This definition highlights the importance of the involvement of all levels of staff and shared values in a true safety culture.

The HSE argues that there is a virtuous circle at work.

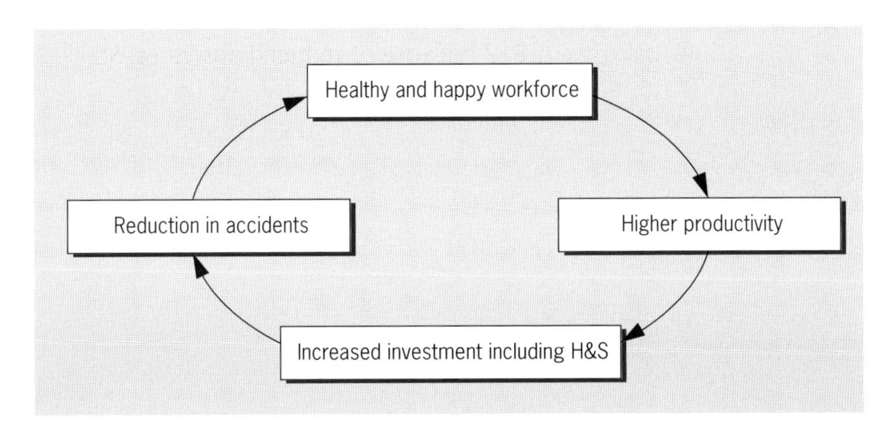


*Figure 1.2 The productivity and safety virtuous circle*

This will involve a substantial shift in attitude for many organisations if they are to move on from compliance towards implementing best practice. For those organisations that make the transition, however, the rewards are well worth the effort.

## Some more good reasons to establish a safe culture

A supportive culture is not just a 'nice to have' or part of the 'feel good factor'. Good reasons to protect health and safety of people at work include:

- laws and regulations
- the statistics for occupational accidents and injury
- the cost of non-compliance
- the number of working days lost.

### The law

The law is very thorough in health and safety. It gives everyone responsibilities and in particular managers. Penalties for non-compliance include criminal prosecution. Did you know that contravening the Health and Safety at Work etc Act 1974 (HSW Act) is a criminal offence? Magistrates are in a position to fine offenders up to £20,000 for breaking health and safety law and in the Crown Courts this could result in responsible individuals or organisations receiving unlimited fines and a jail sentence. Managers can be held responsible for the unsafe behaviour of their employees at work and ignorance of unsafe practices is no defence.

Health and safety inspectors and environmental health officers enforce criminal law and have wide powers. They are in a position to:

- issue improvement or prohibition orders
- enter premises at any reasonable time, day or night
- examine or take possession of any dangerous articles or substances
- investigate accidents or dangerous near misses
- request information or assistance in carrying out their duties
- interview and take written statements from employees
- take out a prosecution.

There are others reasons for establishing a safe culture, which are even more compelling in human terms.

## Accidents, injuries and ill-health

Figures for injuries and fatalities in 2004/5 make sobering reading and indicate the scale of the problem still to be addressed in making the workplace a safer place.

**Fatal injuries**

- 220 workers were killed, a rate of 0.7 per 100,000 workers.
- 361 members of the public were fatally injured.

**Non-fatal injuries**

- 363,000 reportable injuries occurred, according to the Labour Force Survey, a rate of 1,330 per 100,000 workers (2003/04).

**Ill-health**

- 2 million people were suffering from an illness which they believed was caused or made worse by their current or past work.

**Figure 1.3** *Levels of accidents, injuries and occupational ill-health in the UK*

Source: www.hse.gov.uk

Accidents, injury and ill-health are not only distressing for all the individuals concerned. There are costs associated with every incident for an employer, whether that be an insurance claim and higher premiums, the cost of training new staff, time off work, time spent dealing with the injury or accident, costs of repairing equipment and increasingly compensation claims. It is also important to recognise that injuries and accidents can and do happen anywhere.

## Insurance doesn't always cover it

Typically many costs of accidents are not covered by insurance and the organisation still faces a big bill. Insurance may not cover sick pay, damage or loss of products or raw materials, repairs to equipment, delays, overtime working, fines or costs. Claims can also seriously harm your organisation's ability to find affordable insurance cover.

## Working days lost

According to the HSE 28 million working days were lost due to work related ill-health and a further 7 million due to workplace injury.

Some examples of working days lost from the Royal Society for the Prevention of Accidents (RoSPA).

> The average rate of sickness absence from the survey across the whole economy was 4.1 per cent of total working time (9.3 days per employee) with a top rate of about 6 per cent in health, local government etc. and a lowest rate of about 2 per cent in mining and quarrying. Clearly the economic cost for the economy as a whole is massive.

Source: www.rospa.com/occupationalsafety/occupational_health/absence.htm

Any reduction in the level of accident, injury or ill-health can only benefit the productivity of organisations in the UK.

## Activity 2
## Accidents and ill-health

### Objectives

This activity will help you to:

- evaluate your organisation's health and safety culture
- understand the extent to which accidents and ill-health are impacting on your organisation's performance.

### Task

1 How well does your organisation stack up against the evidence for a strong health and safety culture? Evaluate your organisation against the following criteria. Add any others that you can think of at the bottom of the table. Write comments about your organisational culture for each of the points.

| *Evidence of a strong health and safety culture* | *Comments on attitudes of individuals or the culture of the organisation* |
|---|---|
| ☐ Leadership and management with a visible commitment to health and safety | |
| ☐ Long-term strategy with sustained interest and resources to implement measures | |
| ☐ Treating health and safety performance as a business objective | |
| ☐ Commitment to health as a positive contributor to productivity | |

| *Evidence of a strong health and safety culture* | *Comments on attitudes of individuals or the culture of the organisation* |
|---|---|
| ☐ Management practices that put health and safety at the core of managers' role | |
| ☐ A responsiveness to change | |
| ☐ An accident-free workplace | |
| ☐ Working conditions that reduce absence caused by ill-health | |
| ☐ Health and safety practices that are people-oriented and customer-focused | |
| ☐ Standards and clear procedures for working practices | |
| ☐ Communication and co-operation at all levels | |
| ☐ Good questioning attitudes and organisational learning | |
| ☐ A good health and safety record | |

2 Use your own knowledge of the organisation and accident / health records to complete the table below.

| *Types of accidents or ill-health experienced in the organisation* | *Damage to property or adaptations required as a result of accidents or ill-health* | *Types of costs involved e.g. absence, productivity, insured, uninsured* | *Comments on changes in attitude or behaviour required of individuals or the organisation* |
|---|---|---|---|
| | | | |

| *Types of accidents or ill-health experienced in the organisation* | *Damage to property or adaptations required as a result of accidents or ill-health* | *Types of costs involved e.g. absence, productivity, insured, uninsured* | *Comments on changes in attitude or behaviour required of individuals or the organisation* |
|---|---|---|---|
| | | | |

## Feedback

These are both difficult activities to complete. No organisation is perfect. The culture of an organisation and the way it values individuals is difficult to change and needs the support of individuals from the highest levels. Every manager, however can play a part in promoting practices that are not only safe, but also productive.

The important aspect of accident prevention is that if there are accidents occurring then managers need to know why in order to take steps to prevent them (organisational learning). If ill-health is prevalent managers need to make changes to support individuals and prevent occurrences.

If you have difficulty identifying accidents you might think about what could be described as near misses such as, reckless driving, acting the fool, incorrect operation of a machine, or smoking near inflammable liquids. You could also include failure to wear personal protective equipment, not complying with permits-to-work, failing to fit a guard or ignoring warnings.

## What laws do you need to be aware of?

The law should not be a straightjacket for managers. Adopting a responsible, proactive policy based on knowledge of what is required, implementing common sense solutions and monitoring effectiveness are essential components of effective policy implementation.

There are two kinds of legislation that apply in this area. The first is the criminal law which is designed to penalise poor health and safety practice. The second is civil law.

### Criminal law

Criminal Law is concerned with establishing social order and protecting the community as a whole. It gives us a set of rules for peaceful, safe and orderly living. People who commit criminal offences can be prosecuted and if found guilty can be fined, imprisoned, or both.

### Civil law

Civil Law is private law argued out between individuals or between organisations and individuals. It deals with disputes that arise when an individual or business believe that their rights have been infringed in some way.

Under both kinds of law you owe a 'duty of care' to people at work. If you behave in a reasonable way in applying the requirements of the law (both civil and criminal), and follow good management practices of planning, organising, monitoring and controlling, you should be doing enough to comply with the law.

However, some laws place absolute duties on you and failing to comply may result in prosecution.

### The HSW Act and the Management Regulations

The Health and Safety at Work Act 1974 (HSW Act) is the key law in health and safety.

Employers have a general duty to ensure, as far as is reasonably practicable, the health, safety and welfare at work of all their employees.

'Employer' includes you as a manager. It's not just employees that must be protected. The employer also has a duty to people who are not employees, but who may be affected by the organisation's work activities.

What the law requires is what good management and common sense would lead employers to do anyway and that is to look at the risks and put in place sensible measures to tackle them.

### Risk assessment

One of the main requirements for employers is to carry out a risk assessment. The Management of Health and Safety at Work Regulations 1999 (the Management Regulations) add more details to the Act about what employers are required to do to ensure the health, safety and welfare of employees, the community and stakeholders. This can be straightforward in a simple working environment like an office.

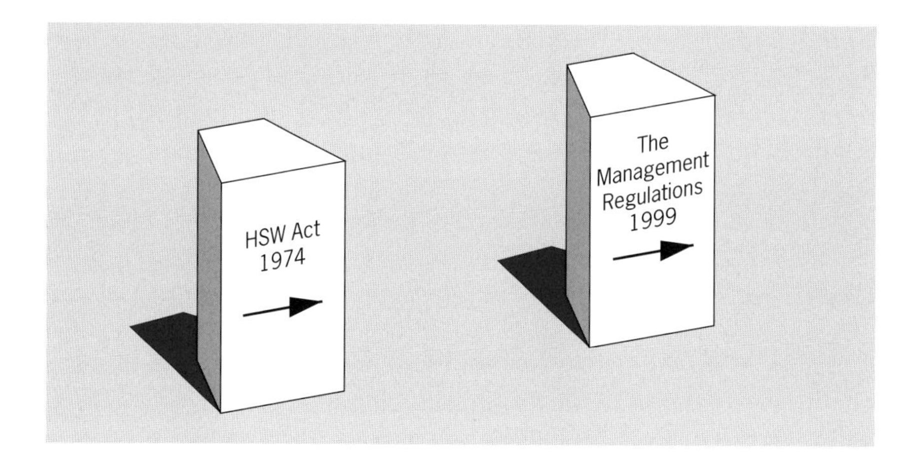


**Figure 1.4** *Two milestones in Health and Safety Legislation*

## Further regulations

Other regulations require action depending on the particular hazards associated with the organisation's activities.

| *Regulations and Acts* | *Basic requirements* |
|---|---|
| Workplace (Health, Safety and Welfare) Regulations 1992 | These cover a wide range of basic health, safety and welfare issues such as accommodation for clothing and facilities for changing, facilities for rest and to eat meals, ventilation, temperature, lighting, cleanliness and waste materials, room dimensions and space, workstations and seating, maintenance for safety, floors and traffic routes, falls and falling objects, doors, gates, walls and windows, sanitary conveniences and washing facilities, drinking water. |

| *Regulations and Acts* | *Basic requirements* |
| --- | --- |
| Health and Safety (Display Screen Equipment) Regulations 1992 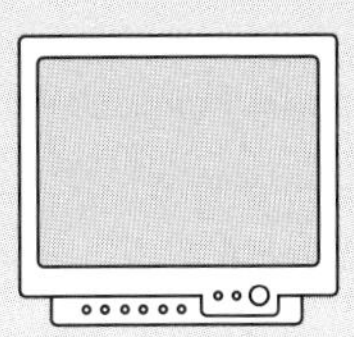 | These regulations set out the requirements for work with Visual Display Units (VDUs) including lighting and periods of use. |
| Personal Protective Equipment at Work Regulations 1992  | These require employers to provide personal protective equipment (PPE) for use at work whenever there are risks to health and safety that cannot be adequately controlled in other ways. It requires that PPE is properly assessed before use to ensure it is suitable, is maintained and stored properly, instructions are provided on its use, and it is used correctly by employees. |
| Provision and Use of Work Equipment Regulations 1998 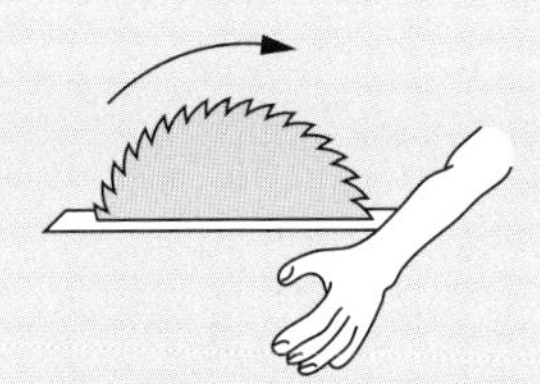 | These regulations require that equipment provided for work is safe. Employers are required to assess all equipment for risks, consider what can be done to reduce the risks, check whether any of these measures are in place and decide whether more needs to be done. If so employers need to do it. This may include assessing whether people have the right equipment for the task, that they have adequate guards, that there are sufficient safety controls and that equipment is maintained. |
| Manual Handling Operations Regulations 1992 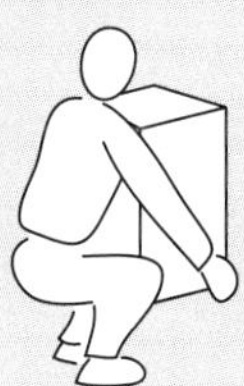 | These cover the moving of objects by hand or bodily force. Employers are required to assess and reduce the risks from manual handling of any objects, from boxes and equipment to people and animals. Employees need to follow appropriate systems of work laid down for their safety by making proper use of equipment provided; co-operate with their employer on health and safety matters; inform the employer if they identify hazardous handling activities and take care to ensure that their activities do not put others at risk. |
| Health and Safety (First Aid) Regulations 1981 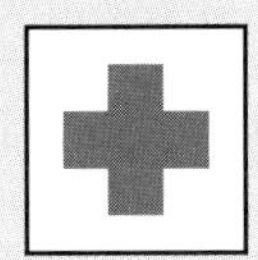 | Employers need to provide adequate and appropriate equipment, facilities and personnel to enable first aid to be given to employees (not necessarily members of the public) if they are injured or become ill at work. What is adequate will depend on the circumstances in the workplace. Employers should carry out an assessment of first aid needs to determine this. |
| The Health and Safety Information for Employees Regulations 1989   | These require employers to display a poster telling employees what they need to know about health and safety. |

| *Regulations and Acts* | *Basic requirements* |
|---|---|
| Employers' Liability (Compulsory Insurance) Act 1969  | This Act requires employers to take out insurance against accidents and ill-health to their employees. 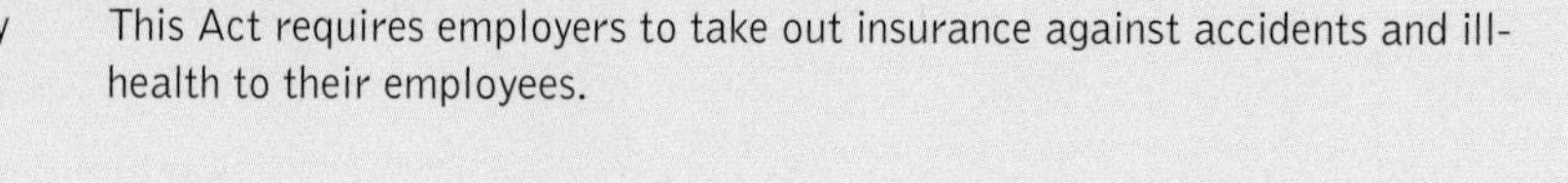 |
| Reporting of Injuries, Diseases and Dangerous Occurrences Regulations 1995 (RIDDOR)  | Employers are required to notify certain occupational occurrences such as death or major injury, an injury resulting in more than three days off work, explosions, chemical leaks, collapse of buildings and heavy lifting equipment and collision with a train. Diseases that are reportable include certain lung diseases, skin diseases, infections such as tetanus and tuberculosis, occupational cancers and poisonings. |
| Noise at Work Regulations 1989 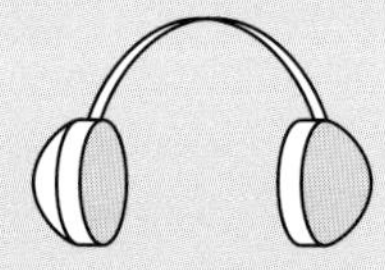 | These require employers to protect employees from hearing damage. |
| Electricity at Work Regulations 1989 | These require people in control of electrical systems to ensure they are safe to use and maintained in a safe condition. |
| Control of Substances Hazardous to Health Regulations 2002 (COSHH)  | These require employers to assess the risks from hazardous substances and take appropriate precaution. |

**Table 1.2** *Key Regulatory signposts*

For certain industries and sectors more specific regulations apply such as:

- Chemicals (Hazard Information and Packaging for Supply) Regulations 2002
- Construction (Design and Management) Regulations 1994
- Gas Safety (Installation and Use) Regulations 1994

- Control of Major Accident Hazards Regulations 1999 (COMAH)
- Dangerous Substances and Explosive Atmospheres Regulations 2002

More details on all these regulations can be found via the Health and Safety Executive (HSE). www.hse.gov.uk

## The Health and Safety Commission and Executive

One of the key provisions of the HSW Act established the Health and Safety Commission (HSC) with responsibility to draft new regulations. The HSC has drafted many of the regulations outlined above.

> **HSC**
> Our job is to protect everyone in Great Britain against risks to health or safety arising out of work activities; to conduct and sponsor research; promote training; provide an information and advisory service; and submit proposals for new or revised regulations and approved codes of practice.

Source: www.hse.gov.uk/aboutus/hsc

The HSE supports the HSC to control and enforce the Acts and Regulations.

> **HSE**
> HSE's job is to help the Health and Safety Commission ensure that risks to people's health and safety from work activities are properly controlled.

Source: www.hse.gov.uk/aboutus/hse/index.htm

## Guidance, Codes and Regulations

Where changes are required the HSC/E can supplement existing laws with Regulations like the ones we have seen above, Approved Codes of Practice or guidance.

### Guidance

The main purposes of guidance are to interpret the law and sometimes this includes showing how European Directives are being met; to help people comply with the law and to give technical advice. These are kept up to date to reflect changes in technologies, risks and events.

## Approved Codes of Practice (ACOP)

These offer examples of good practice and have special legal status. They essentially help to define words such as 'reasonably practicable' or 'suitable and sufficient'. Their legal status is used in law if a company can be found not to have complied with the relevant provisions of the Approved Code of Practice.

## Regulations

Regulations are law, approved by Parliament. They generally allow for some discretion and freedom to control risks in specific circumstances. However some regulations set out actions that must be taken and are absolute. These normally apply when proper control measures are vital.

## In summary

This may seem a bewildering array of Acts and Regulations supplemented by Codes and Guidance. The key things to remember are your duties to employees and others who may be affected by your organisation's activities under the Management Regulations. The main requirement is to assess risks in the workplace and put measures in place to remove or reduce the risks.

## Activity 3
## Relevant Regulations

### Objectives

This activity will help you to:

- get to know the Regulations and guidance related to some common areas of health and safety in the workplace.

### Task

Look at the questions below. Which Regulations cover each of the areas highlighted? You may need to look at the HSE website to get more information.

**Question 1:** Am I entitled to an eyesight text if I work with a computer screen?

**Question 2:** Does an employer have to provide water?

**Question 3:** What breaks are employees entitled to?

**Question 4:** Does an employer have to provide PPE?

**Question 5:** How many first aiders are needed on site?

**Question 6:** What is the maximum / minimum temperature in the workplace?

## Feedback

It may seem strange that you need to know about all of these elements of legislation if you do not have a direct responsibility for health and safety in your organisation. The aim of this activity is to support you in recognising that you have a role, and that there is a lot of detailed information available to help you make a contribution.

**Question 1:**
The regulations that cover VDU work are the Health and Safety (Display Screen Equipment) Regulations 1992. Employers have a duty to ensure the provision of appropriate eye and eyesight tests on request to employees.

**Question 2:**
The Workplace Health, Safety and Welfare Regulations 1992 state that:

'An adequate supply of wholesome drinking water shall be provided for all persons at work in the workplace, that it is readily accessible at suitable places; and conspicuously marked by an appropriate sign where necessary for reasons of health or safety.'

**Question 3:**
The Working Time Regulations 1998 state the following provision for rest breaks at work and time off:

**'Rest breaks at work –** A worker is entitled to an uninterrupted break of 20 minutes when daily working time is more than six hours. It should be a break in working time and should not be taken either at the start, or at the end, of a working day. There are additional regulations that apply to d**aily rest, weekly rest** and y**oung or adolescent workers.'**

**Question 4:**
The relevant regulations are the Personal Protective Equipment at Work Regulations 1992.

'Every employer shall ensure that suitable personal protective equipment is provided to his employees who may be exposed to a risk to their health or safety while at work except where and to the extent that such risk has been adequately controlled by other means which are equally or more effective.'

**Question 5:**
The relevant regulations are the Health and Safety (First Aid) Regulations 1981. In any company, the number and type of first aid personnel would be based on an assessment of the nature, size and remoteness of the organisation.

**Question 6:**
The Workplace (Health, Safety and Welfare) Regulations 1992 and the ACOP deals with the temperature in indoor workplaces and states that:

'During working hours, the temperature in all workplaces inside buildings shall be reasonable. However, the application of the regulation depends on the nature of the workplace. The temperature in workrooms should normally be at least 16 degrees Celsius unless much of the work involves severe physical effort in which case the temperature should be at least 13 degrees Celsius.'

Source: www.hse.gov.uk/contact/faqs/index.htm

## It's your responsibility

The management regulations make an important point about getting organised for health and safety. They require organisations to have arrangements in place to manage health and safety. Part of getting organised is to develop a system of responsibilities and accountabilities – what the HSE calls 'control'. These are not just for managers – everyone has responsibilities in health and safety.

### So what are your responsibilities?

Your responsibilities are defined by the law and can vary from one organisation to another. The following is a summary of your key responsibilities.

1. To maintain buildings, plant, machinery and equipment in a safe condition.
2. To ensure the work environment is safe and free from heath risks and that adequate provisions have been made for employee's welfare at work.
3. To ensure hazardous materials are handled correctly and stored safely.
4. To make risk assessments of significant hazards in the workplace and develop adequate control measures.
5. To report and investigate accidents.
6. To provide information, instruction, training and supervision as necessary to ensure the health and safety of employees.
7. To co-operate with everyone in the workplace in the interest of health and safety.
8. To communicate and consult with relevant groups to promote and develop measures that ensure the health and safety of employees.
9. To plan, implement, monitor and review health and safety controls and keep relevant records of health and safety performance.
10. To use 'competent persons' to assist in health and safety matters.

**Figure 1.5** *A manager's responsibilities under the Management Regulations 1999*

As a manager you should expect some support from others in your health and safety responsibilities.

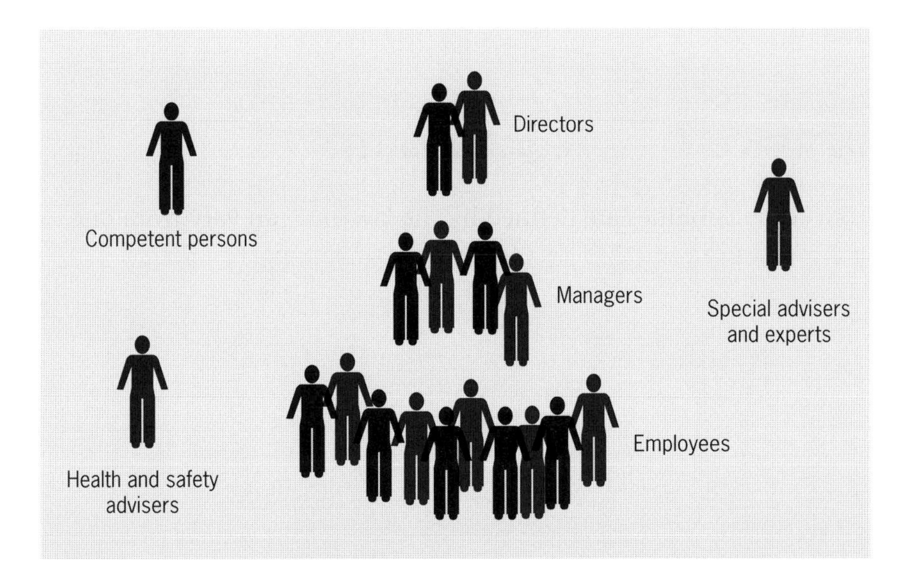


**Figure 1.6** *Range of support for health and safety*

### Directors' responsibilities

The responsibilities are set out in a document from the HSE 'Directors' responsibilities for Health and Safety INDG343 (2001). They can be summarised as follows:

The Board needs to accept formally and publicly their collective and individual roles in providing health and safety leadership. They need to ensure that all their decisions reflect their health and safety intentions as stated in the health and safety policy. Directors also need to ensure that they are aware of health and safety issues and promote the active participation of workers in improving health and safety.

### Employee responsibilities

The duties of employees are:

- To take reasonable care of their own health and safety at work and the health and safety of others who may be affected by what they do or fail to do.
- To co-operate with managers' efforts to carry out legal obligations.
- To follow safety rules, systems and procedures.
- To use work equipment properly.
- To inform managers of health and safety hazards in the workplace.

### Responsibilities of others

The HSW Act outlines the duties to people on full-time, part-time, casual workers, people on work experience, self-employed people and contractors, manufacturers, suppliers, designers, importers of materials used at work and people in control of premises. No one is exempt.

One general duty is imposed on all people, which is not to intentionally interfere with or misuse anything at has been provided at work in the interests of health, safety and welfare.

### Special roles

'Competent persons' have special responsibilities in health and safety. Competent people include, first aiders, health and safety officers or advisers, fire wardens. Competent persons may need special training and qualifications, such as fire wardens or first aiders. However, 'competent' is not necessarily the same as 'qualified'.

Health and safety advisers or officers in an organisation can be particularly helpful in maintaining your responsibilities as a manager. The adviser needs to be both competent and high in status to be able to advise on:

- creating health and safety policies
- carrying out risk assessments
- promoting good health and safety practice
- planning and monitoring health and safety performance
- liaising with external agencies, such as fire authorities, environmental and HSE officers, insurance companies.

They should report directly to directors on matters of policy and have the authority to stop work if it puts people at risk.

### Experts

There are also a wide range of experts available to advise on different types of health and safety problem such as specialist engineers, occupational hygienists, occupational health professionals, ergonomists, physiotherapists and health and safety practitioners.

## Managing performance

Adequate control of health and safety means that people's responsibilities need to be performance based. This means:

- including health and safety in job descriptions
- developing performance targets
- assessing people on health and safety performance
- linking failure to achieve standards or targets in health and safety to disciplinary procedures.

Performance targets may be things such as satisfactorily passing training, or cutting down accident levels or ill-health absences, cutting out certain breaches of procedures or not using set procedures.

## Activity 4
## Your responsibilities audit

### Objectives

This activity will help you to:

- understand your responsibilities in relation to health and safety.

### Task

Go through the following checklist of your health and safety responsibilities and rate your current performance from 1 to 5 – where 1 is poor and 5 is excellent. There are a number of points that are covered in more detail in the next theme. This, however is a good time to think about your current practices and consider where you need to improve your understanding or performance.

- ☐ I maintain buildings, plant, machinery and equipment in a safe condition
- ☐ I ensure the work environment is safe and free from health risks
- ☐ I ensure that adequate provisions have been made for employees' welfare at work
- ☐ I ensure hazardous materials are handled correctly and stored safely
- ☐ I make risk assessments of significant hazards in the workplace
- ☐ I develop adequate control measures based on the assessed risks in the workplace
- ☐ I report and investigate accidents appropriately
- ☐ I provide information, instruction, training and supervision as necessary to ensure the health and safety of employees
- ☐ I co-operate with everyone in the workplace in the interests of health and safety
- ☐ I communicate with relevant groups to promote and develop measures that ensure the health and safety of employees
- ☐ I plan, implement, monitor and review health and safety controls and keep relevant records of health and safety performance
- ☐ I use 'competent' persons to assist in health and safety matters
- ☐ I know who to go to for or where I can find expert support when required
- ☐ I include health and safety targets in performance reviews

**Feedback**

Anything less than 4 or 5 for any statement means that there is room for improvement. Make a note of any issues you need to find out more about. Identify any actions or questions you may have for your health and safety adviser.

# ◆ Recap

This theme begins to establish a framework for a health and safety focus in your organisation. The central tenet of the framework is that health and safety needs to be embedded throughout the organisation with strong leadership support. It needs to be proactive and not just reactive to an accident or a crisis.

**Explore what health and safety covers and its relevance in your working environment**

- Health and safety is wide ranging and covers most aspects of work practice.
- It is both a management practice and a way of benefiting people at work and the organisation itself.

**Identify explicit reasons why you need to bother with it**

- Because the law says so.
- Accidents and ill-health can be reduced with safe working practices.
- There are financial implications to poor health and safety practices, including lost time, insurance premiums and low productivity.

**Determine the laws you need to be aware of**

- Specific laws, regulations and duties apply depending on the nature of your organisation. A range of regulations are illustrated.
- Risk assessment underpins all the regulations and should determine the specific regulations applicable to your organisation.

**Identify your responsibilities as a manager**

- A ten-point summary of your key responsibilities is outlined.
- Managers are primarily responsible for ensuring that a risk assessment is carried out for their area of responsibility and any priorities for action implemented.

- A range of people inside and outside the organisation should be in a position to help managers implement their duty of care to others.

## ➡ More @

**Health and Safety Executive (HSE)** www.hse.gov.uk
The HSE provide a wide range of resources covering specific areas of workplace practice and industry specific guidance. These are available online or from HSE Books.

Guidance relevant to this theme includes:

- 'Health and safety regulation – a short guide' hsc 13 (2003)
- 'Workplace health, safety and welfare – a short guide for managers' INDG 244
- 'Directors' responsibilities for Health and Safety INDG343 (2001)
- 'Health and safety law – what you should know' (1999) This is a short guide for employers and employees.
- 'A guide to measuring health and safety performance' (2001)

**HSE, *Successful health and safety management*, HSG 65 (2000) 2nd Edition, HSE Books**
This book is summarised in the guide: *Managing health and safety – Five steps to success*, INDG 275

**Hughes, P. and Ferrett, E, (2005) 2nd Edition, *Introduction to health and safety at work*, Elsevier Butterworth-Heinemann**
The Introduction to health and safety at work provides a clear outline of all occupational safety and health. It is particularly useful to managers, focusing on their responsibilities in the workplace. It covers the essential elements of health and safety management, the legal framework and risk assessment.

Full references are provided at the end of the book.

# 2 Control and management systems

## Managing health and safety

Management is about organisation. Health and safety management is the same. This theme examines the four fundamental elements that will help you and your organisation to monitor, control and manage health and safety. They are:

- your health and safety policy
- consultation with employees
- risk assessments – reducing risks
- accident investigation and prevention.

With these four elements in place, you have a sound basis for a safe workplace. We will look at the components of each element and how they promote the welfare of everyone in the workplace.

In this theme you will:

- **examine your health and safety policy and its impact on your working practice**
- **identify your role in consulting employees about health and safety**
- **review the hazards and assess the risks in your working environment**
- **understand the basic requirements of accident investigation and prevention.**

## Policy to practice

The HSW Act 1974 makes an important point about getting organised for health and safety. It requires organisations to have arrangements in place to manage health and safety. These arrangements should be identified in the health and safety policy.

A duty is placed on employers to:

> Prepare and as often as may be appropriate revise a written statement of his general policy with respect to the health and safety at work of his employees and the organisation and arrangements for the time being in force for carrying out that policy.

Source: HSW Act 1974 section 2(3)

Any company employing more than five people is legally required to prepare a statement of health and safety policy. This is an important document and one which assists the organisation in identifying the vision and direction it is taking to ensure the health, safety and welfare of its employees.

There are two main elements:

- The first is a general policy statement, which sets out the vision, management commitment, levels of accountabilities and employee responsibilities. This is set out clearly and defined by aims, objectives and targets. It should also indicate when and how the policy will be resourced and reviewed.
- The second element is a plan, which includes an implementation strategy and identifies who will be responsible for implementation.

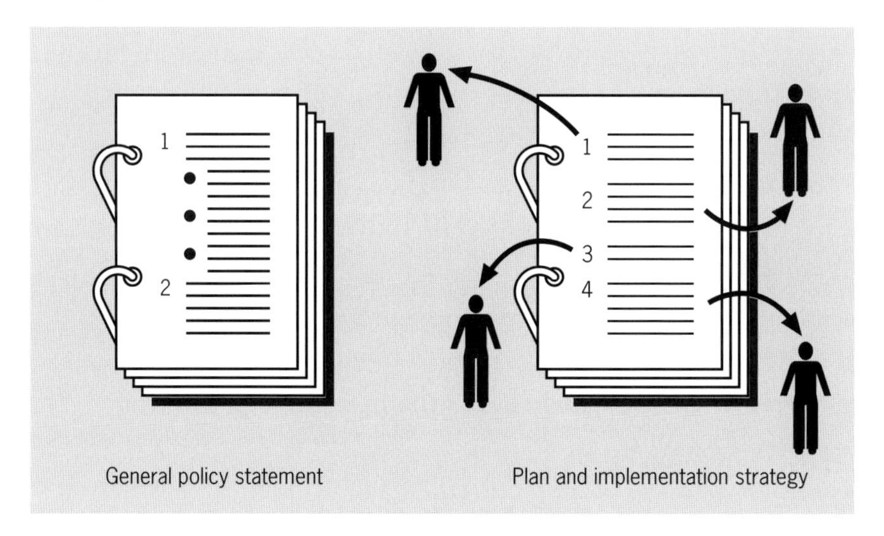


**Figure 2.1** *Components of the health and safety policy*

## Health and safety policy

Typically, a health and safety policy will include the following:

### Aims and objectives

Aims and objectives or 'vision statement' for improving health and safety signed by the Chief Executive – e.g. to reduce accidents and promote good health, comply with the law, protect employees and others in the workplace.

### Responsibilities

- The name of the director with overall responsibility for health and safety.
- The names of other members of the organisation with safety responsibilities.
- The safety responsibilities held by each member.

- The responsibilities of managers for preparing health and safety policies for their departments.
- Relationships with recognised unions.
- Measures to implement consultation with employees on health and safety matters.
- Measures to promote and publicise health and safety and the importance of employees co-operation in achieving and monitoring safe working conditions.
- Sources of specialised and expert responsibilities for safety advice and training.
- Frequency and levels of review of the policy.

A health and safety policy 'vision' statement might look like this from United Utilities:

> Within United Utilities we have a proactive approach to health and safety management, with some of our businesses at the leading edge of company initiatives. We work closely with employees, enforcement agencies, regulators, customers, partners and other stakeholders to develop and promote health and safety and effectively manage our risks. We view this not only as responsible management, but also a commercial opportunity.

Source: www.unitedutilities.com/?OBH=2539

Or like this from Lloyds TSB:

> Lloyds TSB Group plc is committed to the achievement of the highest standards of health, safety and welfare for its employees. These will be achieved by the creation of a positive culture which secures involvement and participation at all levels, sustained by effective communications and the promotion of competence, which enables managers and employees to make a responsible and informed contribution to the health and safety effort.

Source: www.lloydstsb.com/about_ltsb/workplace.asp

## A health and safety implementation plan

The health and safety plan needs to be implemented in context and this can include a range of management systems to support organisations and individuals. This is not just about one off planning, it is about being organised in the long term, so that a health and safety culture is developed. Systems might include:

- an accident reporting and investigation system
- a risk assessment and management system
- training facilities and induction training processes
- provisions for safe maintenance
- measures for the introduction of new machinery, equipment or resources
- monitoring arrangements e.g. inspections
- communication and consultation systems
- planning financial and resource arrangements.

Management systems for health and safety are often summarised as the Four C's – control, competence, co-operation and communication.

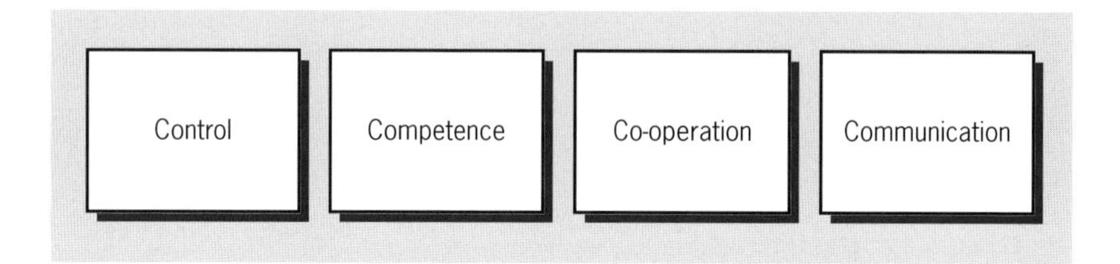


**Figure 2.2** *The Four C's of health and safety management*

## Control

Control measures are those procedures and practical precautions that are in place to prevent accidents and ill-health at work. Exactly what preventative and protective measures you should use in your organisation should be determined by a risk assessment.

Control measures can be physical controls such as guards, safety valves, alarms on machinery or codes of practice which define behaviour. They can be written procedures such as emergency procedures or certificates such as permits to work.

People need to know clearly what they are responsible for and how their duties will be checked. When assessing the adequacy of any control measures in place or when introducing new controls there is a general pattern of risk controls that can be applied.

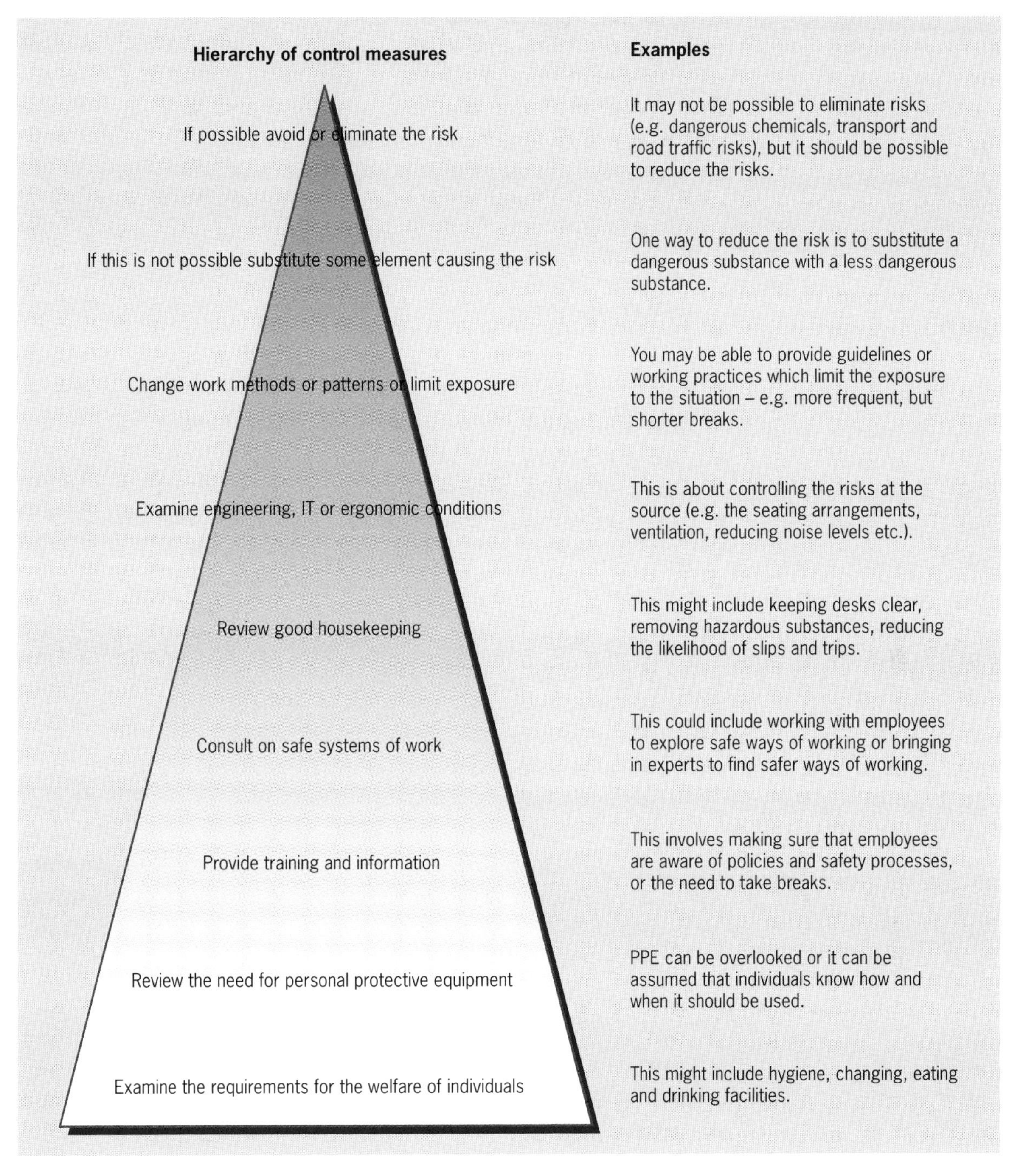


**Figure 2.3** *The hierarchy of control measures*

All risk control measures need to be monitored for their effectiveness and observed to ensure that they are being correctly applied. Monitoring may involve inspections, investigations, consultations, audits, sampling, surveys and workplace tours. Always keep records of monitoring activities so you can show what has been done.

## Competence

Competence for the tasks and activities you undertake includes knowledge and understanding of health and safety issues in practice. Those assigned to specific safety roles also need to be competent for the activities associated with those roles. Employers should provide appropriate training to ensure the safety and welfare of employees. Training might include:

- basic induction training including fire and emergency procedures
- training in specific health and safety aspects of the job, this might be anything from handling heavy complex machinery, operating and moving a laptop to working effectively without stress
- using real life situations to explore how to deal with health and safety risks
- special training may be needed for people in new circumstances, such as pregnant women or young people on work experience
- new training where new equipment is introduced or risks change
- refresher training to ensure that people are up to date and fully conscious of the health and safety practices
- special training for 'competent persons'.

### Co-operation

Health and safety depends on all groups in an organisation being aware of risks and responsibilities. This means understanding what the rules are and how they are enforced. Everyone can and should contribute to the development of safe working practices and consultation and discussion are key to this process. Managers will often set up teams to look at specific problems.

### Communication

Good communication is essential. Information should go up, down and across the organisation in a way that suits the people receiving it. Some examples of communication are posters, safety bulletins, reviews of emergency drills, handbooks with technical information, minutes of safety meetings.

Safety meetings can be a good way of getting information across and seeking ideas. Formal safety committees must meet at least once every three months.

# Activity 5

## The four C's

### Objectives

This activity will help you to:

- examine ways to improve the monitoring and control of safety in your workplace
- use the four C's model to manage health and safety performance.

### Task

Examine improvements you can make to the management of health, safety and welfare in your organisation.

**Control**
An effective management system needs control. People need to know clearly what they are responsible for and how their duties will be checked.

List ways of improving this where you work:

**Competence**
You need to be competent for your work tasks which obviously include health and safety issues. You also need to be competent for any specific safety role. Employers should provide you with appropriate training to ensure the safety culture is successful.

Are you aware of your training and development needs – do you act on them? How could training be used more effectively?

**Co-operation**
Everyone can and should contribute to the development of safe working practices and consultation and discussions are key to this process. Managers will often set up teams to look at specific problems.

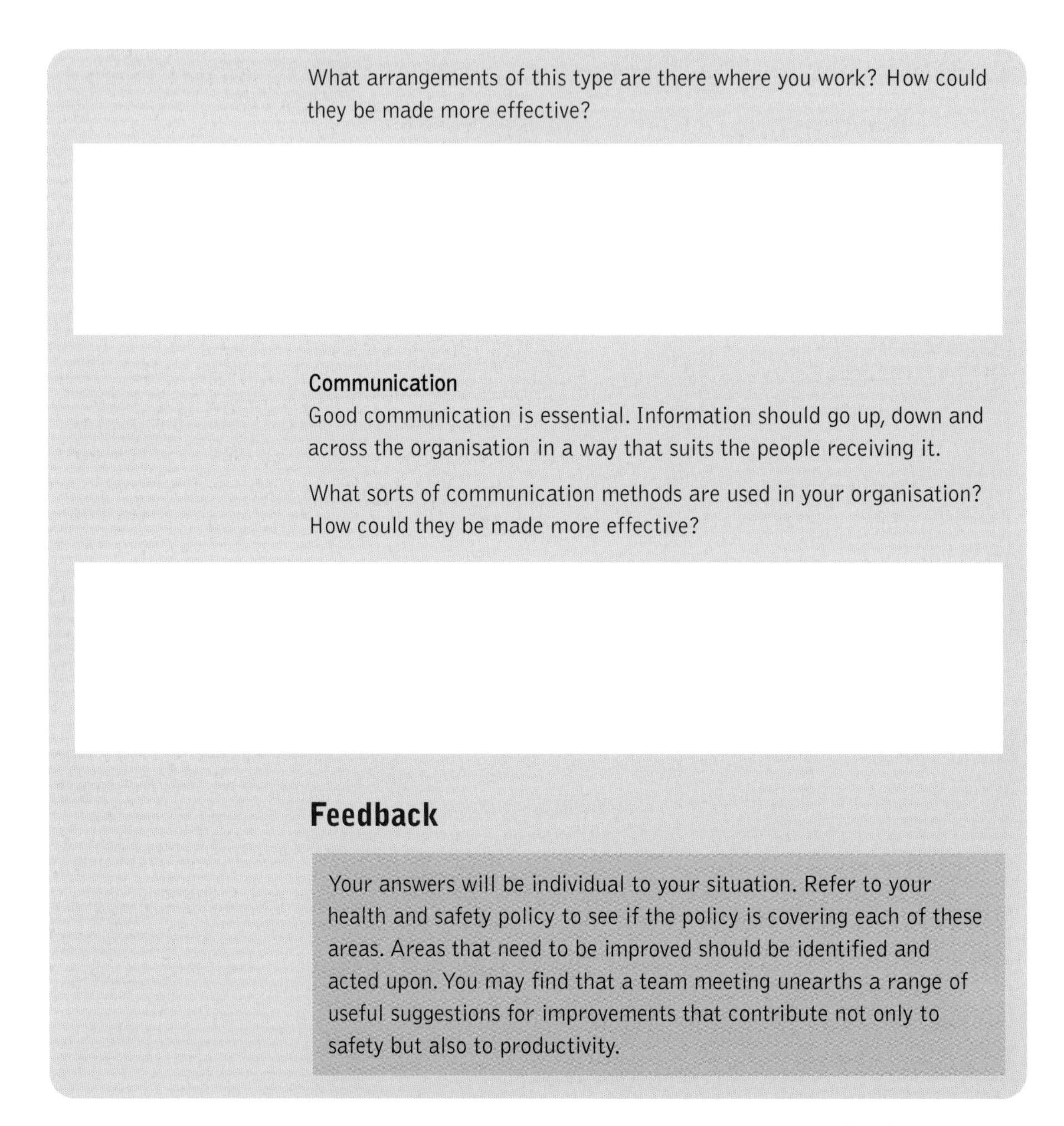

What arrangements of this type are there where you work? How could they be made more effective?

**Communication**

Good communication is essential. Information should go up, down and across the organisation in a way that suits the people receiving it.

What sorts of communication methods are used in your organisation? How could they be made more effective?

## Feedback

Your answers will be individual to your situation. Refer to your health and safety policy to see if the policy is covering each of these areas. Areas that need to be improved should be identified and acted upon. You may find that a team meeting unearths a range of useful suggestions for improvements that contribute not only to safety but also to productivity.

# Consulting employees

## Who to consult

Enshrined in the Management Regulations 1999 there is a requirement to consult employees about health and safety practice. This can be done directly by managers with their employees or via representatives.

## Issues for consultation

Employers must consult on a range of health and safety matters, including:

- any measure at the workplace which may substantially affect their health and safety. For example, changes in procedures, equipment or ways of working;
- the employer's arrangements for getting competent people to help him or her comply with health and safety requirements and evacuation procedures;
- the information employers must give to employees about risks to health and safety and preventative measures;
  - the planning and organising of health and safety training;
  - the health and safety consequences of introducing new technology.

Source: www.hse.gov.uk

For more detailed information about the legal duty to consult, what you should consult on and which system applies to you, see the leaflet *Consulting employees on health and safety – a guide to the law* INDG 232 available from the HSE.

## Benefits of consultation and involvement

Involvement will impact on the commitment of employees to the policies and practices being put into place. Consultation will generate buy-in and is more likely to put health, safety and welfare at the centre of working practices. Involvement in the process will also affect motivation, because all members of the organisation will have a clearer idea of why and how the working practices have been devised. Employees can also have a say in the facilities and accommodation available to them.

The sum total is that if employees are consulted and involved it may be possible to reduce absences due to injury or ill-health and promote a positive working environment with the welfare of all at the centre.

There are a number of stages employees could be involved in:

- **Policy:** deciding on the direction and values associated with the policy.
- **Organising:** defining roles and responsibilities.
- **Planning:** considering options, reviewing working practices and proposing options.
- **Implementing:** implementing practical arrangements and monitoring working practices.
- **Measuring performance:** review.

## Communication

Every organisation is different and you will need to review the way you consult to ensure its effectiveness.

The HSE suggest the following ways of communicating and supporting employees to identify risks and changes to working patterns.

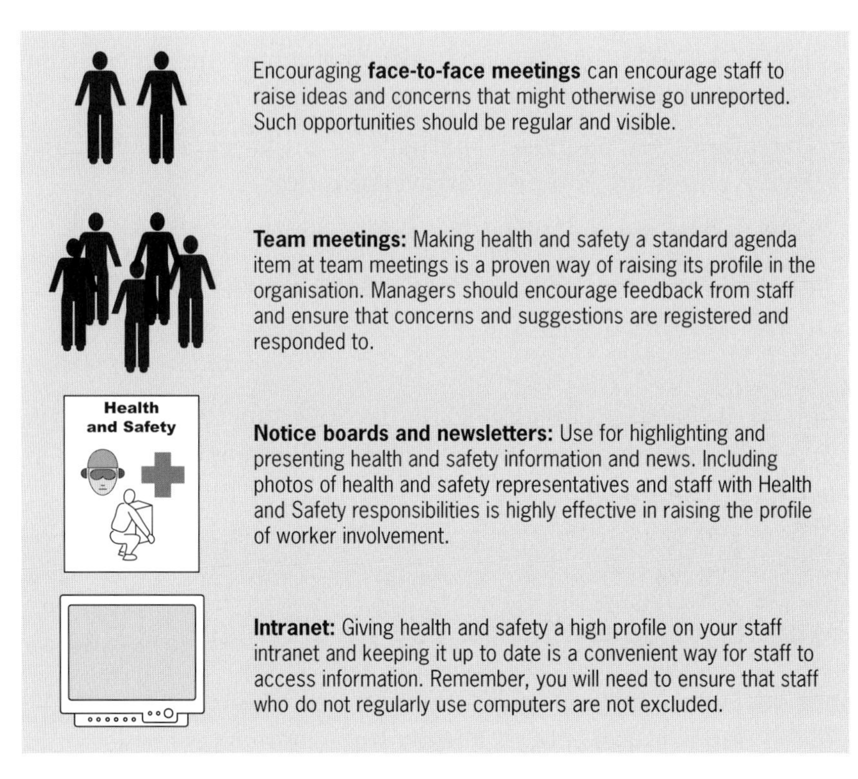


**Figure 2.4** *Effective consultation*

## Activity 6
## Consulting employees

### Objectives

This activity will help you to:

- plan to communicate with employees.

### Task

Your task is to plan a communications session with your team on health and safety. It could cover:

- any change which may substantially affect their health and safety at work, for example in procedures, equipment or ways of working

- the information that employees must be given on the likely risks and dangers arising from their work, measures to reduce or get rid of these risks and what they should do if they have to deal with a risk or danger
- the planning of health and safety training
- the health and safety consequences of introducing new technology.

Consider when and where it would be appropriate to hold the session, how you will run it, what props you would use, and how to achieve an effective two-way dialogue. You will also need to think about sources of further information and how the actions from the meeting with be carried out and followed up.

If possible, carry out the communications session and write down the key results of the meeting.

### Feedback

This is an important element of your responsibility for health and safety. It should not only generate ideas for good practice and improvements, but it will also demonstrate your commitment to the health, safety and welfare of your employees.

## Hazards and risk assessment

Risk assessment is the basis for most UK and European law on health and safety. Rather than prescribing specific practices, the governments of Europe have indicated that organisations should identify and respond to the risks which arise in the specific circumstances of their activities. They have then legislated, with Regulations, to control practices in each of the circumstances.

### Risk assessment

Ridley (2004) describes a risk assessment as:

> a means whereby an employer can manage properly the risks faced by his employees and ensure that their health and safety is not put at risk while at work.

Source: Ridley (2004)

Risk assessment is basically a five-stage process:

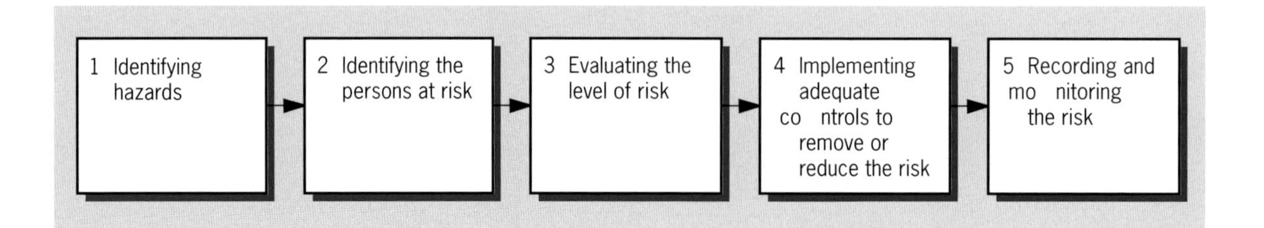


**Figure 2.5** *The five stages of risk assessment*

These are expanded upon in the Management Regulations, which place specific responsibilities on employers to:

- carry out 'suitable and sufficient' risk assessments of hazards identified
- decide what is 'suitable and sufficient' in the light of their operating circumstances
- cover in the assessment all equipment, both existing and new and materials and substances
- give priority to protecting the whole work force rather than individuals
- consider any risks from their operations that may affect non-employees
- appoint an assessor to carry out the assessments. The assessor should have a knowledge of the work processes, the legislation and health and safety standards for the industry
- give the assessor time to carry out the assessments during working hours
- record the results of the risk assessment.

A useful text for managers is *Five steps to risk assessment* INDG 163 (2003) published by HSE.

A risk assessment carried out in the work place needs to be systematic. It could be quantitative – that is attempting to measure the probability and possible severity of the risk and assigning a numerical value. This kind of risk assessment is more common where a problem could be very serious. More normally, the risk assessment is qualitative and rated as high, medium or low risk.

## Stage 1 Identifying hazards

Only significant hazards, which could result in serious harm to people, should be identified. A tour of the area is usually essential, as are a review of the accident, incident and ill-health records and any safety inspections, surveys or audits. It may be necessary to review codes of practice and manufacturers guidelines.

## Hazards checklist

Some of the most common hazards areas include:

- ☐ **equipment, electrical or mechanical malfunctions or poor working practice** – impact, abrasions, cutting, entanglement
- ☐ **transport** – mechanical handling, people and vehicle collisions, operation of forklift trucks, parking and reversing
- ☐ **as a result of access** – slips, falls, trips, moving objects, obstructions, working at height, confined spaces, excavations
- ☐ **manual handling** – poor positioning or posture, supporting or lifting people, carrying heavy loads, inappropriate use of mechanical handling equipment
- ☐ **chemicals** – dust, fumes gases, toxicity, irritants, corrosion, carcinogens
- ☐ **fire and explosion** – flammable materials, gases or liquids, explosions, means of escape and alarms for specific purposes
- ☐ **radiation** – ionizing and non-ionizing
- ☐ **biological** – bacterial, viral or fungal
- ☐ **environmental factors** – noise, vibration, light, ventilation, temperature, overcrowding
- ☐ **human factors** – individual not suitable for the task, long hours, violence to staff, unsafe behaviour, stress, pregnant/nursing women, young people
- ☐ **other factors** – Poor maintenance, lack of supervision, lack of training, lack of information and instruction, unsafe systems.

**Figure 2.6** *Hazards checklist*

Source: adapted from p 69 Hughes and Ferrett (2005)

### Stage 2 Identifying the persons at risk

People who work in the organisation are at the most obvious risk from your organisation's activities. However, it is important to consider anyone else who may be impacted by your activities and this could include the neighbouring communities. The impact could be anything from increased traffic on the roads to environmental waste.

The more people and the more seriously people are at risk the higher this risk moves up the priority ladder.

### Stage 3 Evaluating the level of risk

Risks identified in the risk assessment may need to be rectified immediately, such as replacing guards on equipment or using appropriate PPE. The goal is to reduce the levels of potential risks. This is achieved by estimating the severity of the risk and rating it against the likelihood of the risk. The higher the combined severity and likelihood ratings the higher this should appear on the timetable for action.

Hazards need to be balanced against who is likely to be harmed and against the existing controls in place to provide a true reflection of the likelihood of the risk arising. One way of presenting this is as follows:

| *Hazards* | *People at risk* | *Existing controls* | *Controls required* | *Priority level* |
|---|---|---|---|---|
| Fire – electrical faults from dense use of computer and electronic equipment | All employees and visitors, staff in the offices below | Fire and emergency procedures including extinguishers, fire detectors and alarm systems<br>Electrical checks<br>Employees trained in good housekeeping | Discussions with the insurer on the provision of fire safety equipment<br>Further checks on electrical equipment brought into the building | Medium |

**Table 2.1** *Register of hazards and controls*

This style of layout gives a clear indication of the controls in place and any suggestions for further controls, as well as a priority rating.

### Stage 4 Implementing adequate controls to remove or reduce the risk

Based on the hazards, the persons affected and the priority assigned to each risk it should be possible to generate a timetable for rectifying, removing or reducing the risk. As identified earlier in this theme there is a broad hierarchy of control which should be applied with the objective being to eliminate or reduce risks. It is important to maintain a continuous programme of risk improvement. This may mean tackling some of the low priority risks at the same time as the high priority ones.

### Stage 5 Recording and monitoring the risk

It is vital to keep a record of the risk assessment and any new control measures that have been implemented to reduce the impact. This written record provides evidence to inspectors, and directors of the company of compliance with the law. It is also useful as evidence if a criminal or civil action is taken against the company. Employees are entitled to see the risk assessment and it should be kept with the health and safety policy.

Risk controls should be reviewed regularly and actions and priorities monitored for use and effectiveness. Review and revisions may be required due to changes in personnel, equipment, legislation or processes. An accident may also prompt a review of the risk assessment to try to establish if another similar accident could be avoided. Your health and safety policy may stipulate how often your risk assessment must be reviewed and how actions resulting from the review will be monitored.

In addition to a general risk assessment it may be necessary to carry out specific risk assessments for areas specific to your activities, such as display screen equipment, PPE, transport, dangerous substances, etc.

## Activity 7
## A picture of performance

### Objectives

This activity will help you to:

- create a picture of your overall performance and monitoring of health and safety
- identify measures for prevention and control of hazards.

### Task

Measuring performance will help you to assess the overall performance of your team and area of responsibility for health and safety. You should seek to answer the following questions.

- Where are you now relative to your overall health and safety aims and objectives as stated in the health and safety policy?
- What measures do you have in place to control hazards?
- What identified hazards and risks are outstanding from the last risk assessment? Could any hazards be eliminated?
- What measures do you have in place to prevent accidents and ill-health?
- Is your organisation getting better or worse over time?
- Is your management of health and safety effective (doing the right things)?
- Is your management of health and safety reliable (doing things right consistently)?
- Does your culture support health and safety, particularly in the face of competing demands?

### Feedback

These questions should be asked not only at the highest level but also across the organisation. The aim should be to provide a complete picture of the organisation's health and safety performance.

# Accident investigation and prevention

## What are accidents?

The dominoes below represent the sequence of events that lead to an accident. If one of the dominoes falls there is likely to be a chain reaction. Practical intervention is required to separate the dominoes.

> An accident is not a single event; it is the result of a series of linked causes.

Source: Ridley 2004

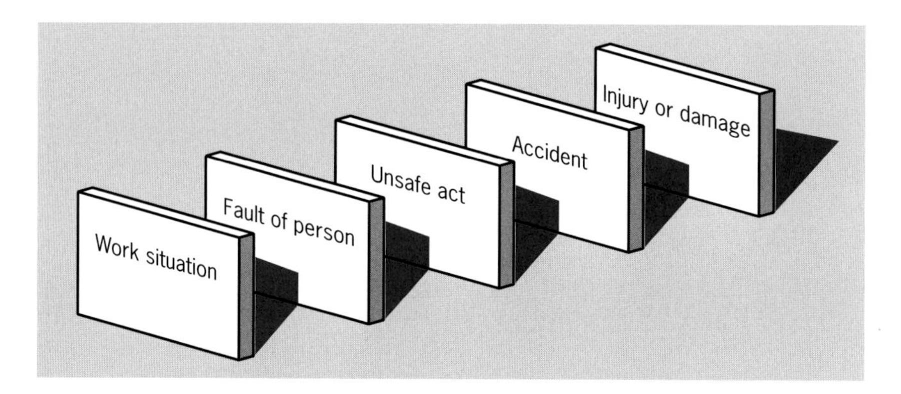


**Figure 2.7** *Heinrich's domino theory*

A manager's responsibility is to ensure that the work situation has adequate management controls, suitable standards for performance and operational equipment. In turn an individual and their manager will need to ensure that the operator has appropriate levels of skill and knowledge, the capability to carry out the work and the motivation to do it properly. The unsafe act may involve not following agreed methods of work, taking short cuts or removing safety equipment or PPE.

## Investigating accidents

The HSE propose a general four step model for investigating accidents and the four steps are:

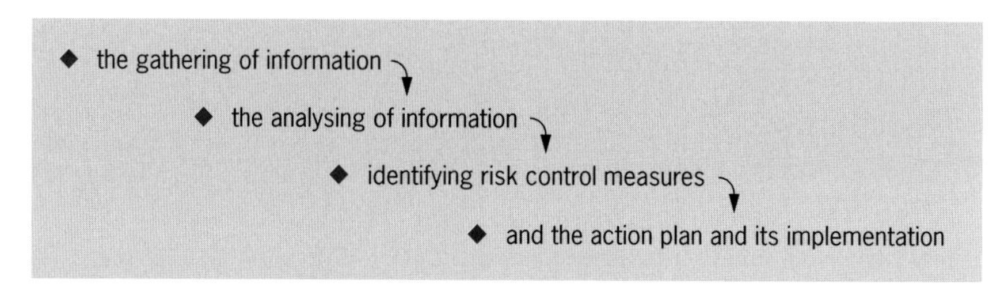


**Figure 2.8** *The four step accident investigation model*

This is explained in more detail in *Investigating accidents and incidents – a workbook for employers, unions, safety representatives and safety professionals* HSG245 from HSE Books, PO Box 1999, Sudbury, Suffolk CO10 2WA. Another useful reference guide is *RIDDOR explained* HSE31.

A key element of accident reporting and investigation is the gathering of information. You will need to record immediately:

- the date, time and place of the event;
- personal details of those involved; and
- a brief description of the nature of the event or disease.

You could choose to keep your records by writing a report and keeping it in a file or on a computer, or you could keep a written log noting the time and date of each entry. These records will form the basis for any evidence and should be sufficient as a record for most purposes.

## Accident Investigation Challenge

The Royal Society for the Prevention of Accidents (RoSPA) has devised the Accident Investigation Challenge to highlight the importance of accident investigation as the basis for learning and preventing accidents. The objectives are to prevent accidents happening and to determine the cause of accidents to prevent reoccurrence.

The following prompt list has been developed by RoSPA to help organisations check out where they are now:

1 **Commitment to learning**

Does everyone understand and accept that the organisation is fully committed to learning from its health and safety failures? (For example, that it is more interested in learning lessons which can help it improve its management of OS&H than it is in merely allocating blame.)

2 **Reporting**

Does every employee feel obliged and empowered to report promptly and accurately all accidents, incidents and safety significant issues, which come to their attention? (For example, are they actively encouraged to report errors and safety failures? Can they be confident that they will be valued for doing so? Do health and safety performance targets, for example, tend to act as a disincentive to reporting accidents and incidents?)

3 **Scaling and terms of reference**

Are there adequate and suitable processes and criteria (e.g. risk/consequence or learning potential) in place to enable the organisation to decide on the scale and depth of investigation and to draw up initial terms of reference? (Does the organisation simply scale its investigation response according to the severity of injury or does it consider the safety significance of the each accident or incident and its potential for improving safety in the future?)

4 **Team based approaches**

To what extent does the organisation adopt an open, team based approach to investigation, with effective involvement of operative level employees, safety representatives, and supervisors, drawing on their practical knowledge and providing opportunities for them to learn more about safety and become champions for necessary safety change? (Is the team led by a manager with appropriate seniority?)

5 **Training, guidance and support**

Have all team members received necessary training and guidance to enable them to play their part effectively in the investigation process, for example, training in interview techniques? Is practical guidance and technical support available to the team from qualified H&S professionals?

6 **§Information gathering**

How adequate are existing procedures in enabling investigators to gather necessary data following accidents and incidents – including for example: securing the scene, gathering essential physical and documentary evidence, taking photographs (for example, using digital cameras), interviewing witnesses etc?

7 **Use of structured methods**

Does the organisation make use, as appropriate, of structured methods to enable it to identify the circumstances of which the accident or incident is the outcome? Does it use such methods to help it integrate evidence, generate and test hypotheses and reach conclusions so it can make recommendations?

8 **Immediate and underlying causes**

Do investigations seek to identify and discriminate between immediate and underlying causes? Is there a clear link between the outcome of investigations and revision of risk assessments, for example, to establish if and why risk assessments for the activities concerned were inadequate, had not been properly implemented or had been allowed to degrade.

9 **Communication and closure**

Are there effective means in place to communicate conclusions back to stakeholders and to track closure? Is the implementation of recommendations managed to an agreed timetable with reporting back to the investigation team?

10 **Reviewing investigation capability**

Does the organisation undertake a periodic review of the adequacy of its approach to investigation with a view to improving its capability to learn lessons from accidents, incidents and occupational safety and health problems and to embed these lessons in 'the corporate memory'?

Source: Learning from safety failure, www.rospa.com/occupationalsafety/learning/index.htm

Organisations that have difficulty in providing robust responses to these questions are strongly urged to review their current approach to learning from accidents.

## Activity 8
## Accident investigation challenge

### Objectives

This activity will help you to:

- understand the process for accident investigation in your context
- consider ways the process could be improved.

### Task

This task asks you to consider your organisation's overall attitude to accidents and occupational ill-health and if possible your response to an incident. Either, consider an incident which resulted in injury or prolonged ill-health that has occurred in your workplace or referring to your health and safety policy and risk assessment consider a potential incident that could occur. Answer the questions from the RoSPA Challenge above.

### Feedback

These are challenging questions that you may like to discuss with other managers in your organisation. Your answers may not always be positive; that is the nature of organisations responding to difficulties. Think about ways managers can work together to learn from accidents and prevent them occurring in the first place.

# ◆ Recap

This theme explores the structures and systems that can be used to promote safety in the workplace.

**Examine your health and safety policy and its impact on your working practice**

- Health and safety policies should be designed to set the agenda for safe working practices offering clear guidance with clear responsibilities for action.
- The Four Cs of health and safety management systems, Control, Competence, Co-operation and Communication are the foundation for good working practices.

**Identify your role in consulting employees about health and safety**

- Employers are obliged to consult employees about health and safety issues in the workplace.
- However there are compelling reasons why this is to the advantage of everyone involved.
- Communicating and supporting employees to identify risks and changes to working patterns is not only likely to generate a safer workplace, but also a more productive one.

**Review the hazards and assess the risks in your working environment**

- There are five stages in the process of identifying hazards and assessing risks: Identifying hazards; Identifying the persons at risk; Evaluating the level of risk; Implementing adequate controls to remove or reduce the risk; and Recording and monitoring the risk.

**Understand the basic requirements of accident investigation and prevention**

- Accident and investigation is as much about learning from mistakes as it is about recording the facts for investigation or insurance purposes.
- Prevention of accidents is part of an underlying organisational culture for safety.

# ➤➤ More @

**Health and Safety Executive (HSE) www.hse.gov.uk**
The HSE provide a wide range of resources covering specific areas of workplace practice and industry specific guidance. These are available online or from HSE Books.

Guidance relevant to this theme includes:

- 'Consulting Employees on Health and Safety: a Guide to the Law' INDG 232 (2002)
- 'Five steps to risk assessment' INDG 163 (2003)
- *'RIDDOR explained'* HSE 31 (1999, reprinted 2004)
- HSE, *Investigating accidents and incidents – a workbook for employers, unions, safety representatives and safety professionals* HSG245 from HSE Books, PO Box 1999, Sudbury, Suffolk CO10 2WA.

**Hughes, P. and Ferrett, E, (2005) 2nd Edition, *Introduction to health and safety at work*, Elsevier Butterworth-Heinemann**
The Introduction to health and safety at work provides a clear outline of all occupational safety and health. It is particularly useful to managers, focusing on their responsibilities in the workplace. It covers the essential elements of health and safety management, the legal framework and risk assessment.

**Ridley, J. (2004 )3rd Edition, *Health and safety in brief*, Elsevier Butterworth-Heinemann**
This is a concise source of essential legal information and best practice for managers. Ridley attempts to take some of the mystique out of health and safety in the workplace to explain the hows, whys, and wherefores of keeping on the right side of health and safety laws. It covers law, management, occupational health and safety technology.

**Royal Society for the Prevention of Accidents (RoSPA) www.rospa.com**
There are some very useful ideas and case studies to support the prevention of accidents on this website including fact sheets on Road Safety, Safety Education, Home Safety, Water and Leisure Safety, Occupational Safety, Product Safety, Play Safety and Driver and Fleet Solutions.

# 3 Facilities management and ergonomics

## Space to work

Your facilities are your buildings, office or workspace. The more efficiently they are run; the more benefit to your organisation. Making the most of your facilities involves looking at the range of people working in the environment, people who visit your place of work and factors associated with access, size and cost.

Space to work is also explored in this theme. The importance of space is becoming increasingly recognised. Over the course of a full year, the average full-time worker spends about 30 percent of their waking time in their place of work. We look at ergonomics; the factors in the working environment which define levels of comfort, influence behaviour and can contribute to productivity or lead to accidents and ill-health.

There are legal implications for setting up premises and equipment for business. This theme looks at issues like insurance, access, planning and hiring out buildings.

The facilities you use to operate should offer appropriate levels of security to individuals working or visiting the environment. This includes protecting contents, storing materials and providing security from malicious attack. You will also need to consider emergency procedures on the premises.

In this theme you will:

- **examine the range and effectiveness of your facilities**
- **understand the impact of ergonomics, space to work and the human factors that influence behaviour**
- **review the legal obligations and insurance requirements including location, planning access, size, hiring out and cost factors**
- **develop an understanding of security and safety implications for employees and premises.**

## Effective use of space

Making the most of your workspace makes financial, social and common sense. The influence of your workspace on your employee's work and the image of your organisation on people from outside are easy to underestimate. Here, we look at the use of your facilities in relation to the work that you do, the range of people who come into contact with your facilities and the efficiency of the space in terms of planning, access, size and cost.

We will examine:

- what the space is used for
- who uses the space
- organisational image
- size, location and access.

## What is the space used for?

The layout of a workspace, whether that be a building, an agricultural setting, an office, a shop, a factory, a school or a hospital refers to the space available for employees to meet people, use equipment, machinery or goods. A well-planned layout should make a significant contribution to the efficiency of the process or operation being carried out.

For most managers in established organisations, this will probably have been broadly worked out for you. It is worth however, identifying the features of the layout which currently contribute to an efficient workflow – you may then be able to identify elements where the workflow is interrupted or could be improved. A manager with a new workspace to plan will face different problems. A starting point for both, however, is to review the process and tasks carried out.

You may want to review:

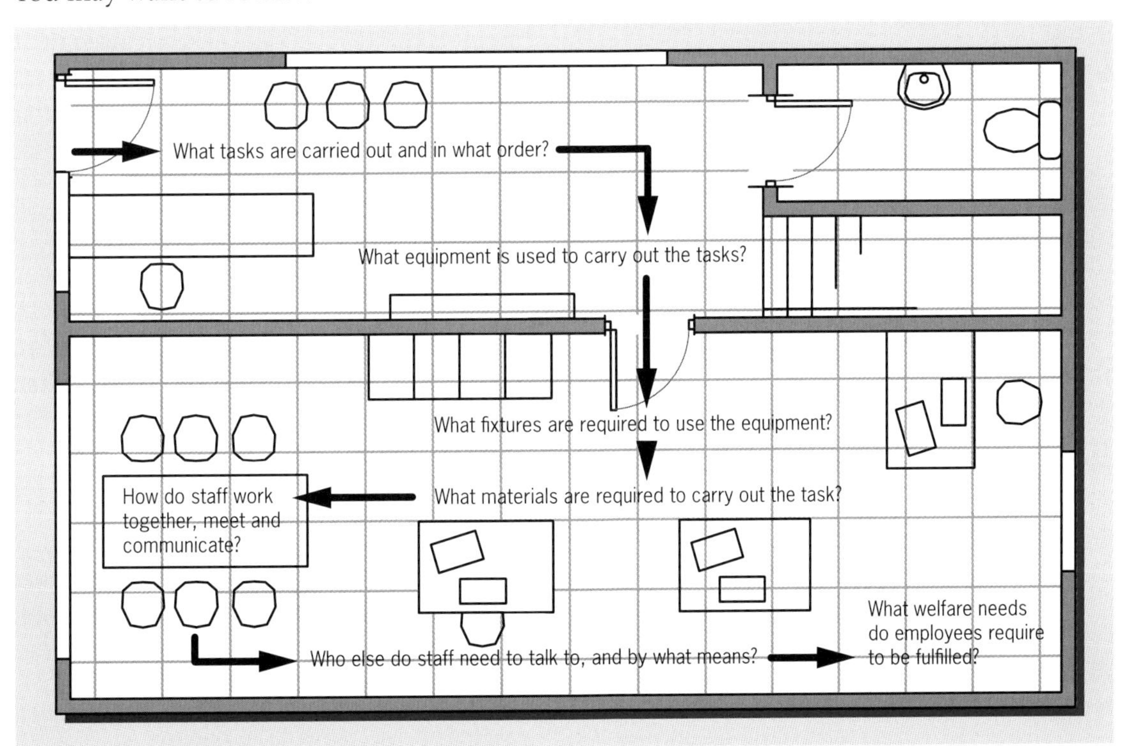


**Figure 3.1** *Reviewing work processes and tasks*

A good workspace layout will ensure that your processes move smoothly from one stage to the next with a minimum of lost time. It is rarely possible to be completely logical about the space required and the space available, but the aim is to minimise the compromises and make the most of what space you have.

## Who uses the space?

Once you have determined what the space is needed for, you can review who uses the space. As we have seen the majority of people spend about 30 percent of their waking hours in their place of work and some significantly more. The environment they work in will play a significant part in their:

- productivity
- security
- ability to communicate effectively
- levels of comfort
- well-being
- health
- safety.

It becomes very apparent how important space to work becomes.

### Productivity, security and communications

Productivity can be improved by attending to the workflows and processes required to complete tasks. It will also be affected by motivation. You may need to consider whether employees basic or welfare needs are being met. It may be worth asking them. Do they have any concerns for their safety? If they are feeling secure, are they able to work effectively as a team and communicate with others easily?

### Comfort, well-being, health and safety

The way space is arranged will support comfort and well-being, or lead to insecurity and ill-health. Think about the environments you and your colleagues work in and examine the factors that promote comfort, health and safety and those that don't. Again ask your team for their comments.

### Space for customers, visitors and the public

Many organisations must also allow for customers on the premises. The requirements of customers, visitors, employees and the public need to be taken into consideration in facilities management.

There are often conflicts of interest in the division of space. Customers will not want to be faced by a noisy back room environment typical of a supermarket warehouse. IT departments will not want to be interrupted by listening to the calls of a telesales team. Separation of space is one of the key areas of facilities management, but by separating people you may be putting up barriers to effective communication between teams and some compromises may be required.

## Organisational image

The image portrayed by an organisation is dependent on many aspects of the way it presents itself. The facilities it uses may be a significant element in the image, for instance shops in retail operations or taxis in a taxi firm. The image affects the attitudes that people hold towards the organisation. A person's attitude might, for example affect whether they would apply for a job or buy goods. These attitudes are formed from what someone knows about an organisation and how they feel about it. The facilities used by the organisation are likely to bring out a gut reaction and could affect the way an individual or customer deals with you or whether you can recruit and retain the right staff.

By maintaining and presenting facilities appropriately an organisation can:

- make the selling process easier – people are more comfortable buying from an organisation that looks proud of its image and its people
- contribute to a positive brand image
- improve morale among employees
- recruit and retain staff.

## Size, location and access

Obviously, we would all like room to spread out. But space costs money. Too little space leads to overcrowding.

Size and location decisions are fundamental to the efficiency of an organisation. They can have an impact on the quality and quantity of relevant employees, environmental factors, transportation costs, proximity to customers, availability of transport networks for employees. Location will also involve consideration of local community reactions, density of surrounding buildings, availability of parking areas, traffic control systems, heavy vehicle access and vulnerability of the public to involvement in major accidents.

### Access issues

Access to premises is vital to facilitate entry for employees, customers and visitors, including a diverse range of people with disabilities or impairments.

Overall, effective access is achieved by:

- a level approach to the building or from the car park
- on-site car parking and setting down
- ramped access with handrails where required
- clear signage especially for visitors and customers.

### Floors and traffic routes

Corridors, stairs and pathways are a key element of any workplace, allowing free movement around the facilities. They are effective if they connect people who need to work together and they are well signposted. Planning of the office or production space needs to include all elements of the tasks and workflows that happen in an organisation.

## Activity 9

## Space issues

### Objectives

This activity will help you to:

- assess the effectiveness of the use of space in an area for which you have responsibility
- suggest improvements for use of space to make it more effective.

### Task

You will have a number of priorities for the use of space in your work environment. Against the following categories, write some notes on the effectiveness of and the constraints which limit your use of space. You may be able to suggest some improvements that could be made. An example is provided.

| *Priorities for use of space* | *Effectiveness of current use* | *Constraints* | *Improvements* |
|---|---|---|---|
| Staff work processes | | | |

| *Priorities for use of space* | *Effectiveness of current use* | *Constraints* | *Improvements* |
|---|---|---|---|
| Staff welfare needs | | | |
| Customer areas | | | |
| Meeting areas | | | |
| Materials and equipment usage, safety and security | | | |

## Feedback

Too often design of space centres on where computers are sited and who has a window seat. The welfare and privacy of staff and customers must also be considered as a high priority. How are they being catered for?

Staff work processes contribute to productivity when they are well arranged, including facilities to communicate effectively. Could processes be improved by moving people around to improve communications? Could equipment be sited more effectively to ease the flow of materials or processes? There may be little choice about some of the arrangements due to constraints.

It is always worth asking those that use the space on a day-to-day basis how effectively it is being used and any improvements that could be made.

## The human factor

Effective use of space, equipment and materials are all dependent on humans. But as we all know ...

> To err is human.

Source: Alexander Pope (Poet)

Human error is thought to be the main reason for approximately 80% of accidents, which suggests a great deal more can be done to prevent them.

### What is ergonomics?

HSE has defined human factors (also known as Ergonomics) as:

> the environmental, organisational and job factors, and human and individual characteristics which influence behaviour at work.

Source: www.hse.gov.uk/humanfactors/index.htm

Improvements in the factors which influence behaviour at work can improve health and safety, firstly by reducing the number of accidents and cases of ill-health at work. And secondly, by providing considerable benefits for business including reducing the costs associated with such incidents and increasing efficiency.

### Who is at risk?

It's not just people lifting heavy goods or people working on construction sites, although these are the areas where a great many incidents and accidents occur.

Ergonomics and the workspace we use impact on everyone. Have you ever found yourself with eyestrain from focusing on a computer screen too long? Or have you lost concentration and provided a bad service or produced low quality goods because you haven't taken a break? Or got backache from a poor seating arrangement? This is an issue that impacts on you, the organisation and the economy.

## Why do the problems arise?

Problems arise because as humans we are prone to forgetting to take care of ourselves, working practices are ill thought out or the work place is badly designed. The sort of issues that human factors/ ergonomics can make a significant contribution to tackling are outlined by the HSE as follows:

- Prevention of musculoskeletal disorders and manual handling injuries
- Management of work-related stress
- Preventing falls from height
- Preventing slips and trips
- Preventing workplace transport accidents
- Human factors in design (e.g. vehicle cab or workspace design, etc.)
- The design and effectiveness of procedures
- Human reliability – human error and systems failures e.g. maintenance error
- Staffing levels and workload
- Fatigue from working patterns – shift work and overtime
- Training and competence.

Source: www.hse.gov.uk/humanfactors/live.htm

## How do we know we have a problem?

There are two aspects to knowing that you have a problem. The first is that health problems are evident. The second is that productivity and efficiency are reduced. Health problems show up in different ways. There could be an increase in accidents, days off through ill-health, reports of aches and pains and employee complaints. Productivity problems show up in poor product or service quality, high waste and low output.

It is your role as a manager to pick up on these issues and try to work out the root causes. They may appear in the course of a health and safety risk assessment or when you are looking at welfare issues. Employees may bring up issues. By whatever means they are raised, they are significant, demand attention and risks should be assessed. Improvements can be made by looking at:

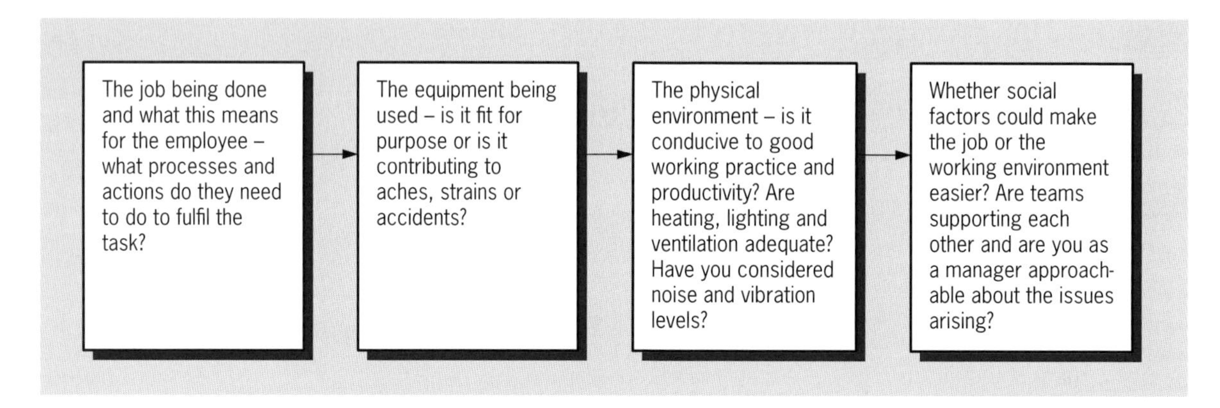


**Figure 3.2** *Improving process flows*

## Why do we need to take action?

We need to take action to prevent accidents and ill-health, to improve productivity and because the law says so.

There are no specific regulations on human factors. However human factors/ergonomics are an element of compliance for some of the regulations and guidance. Take for example:

- Control of Major Accident Hazards Regulations 1999 (COMAH);
- Railways (Safety Critical Work) Regulations 1994;
- The Manual Handling Operations Regulations 1992 (as amended);
- The Health & Safety (Display Screen Equipment) Regulations 1992 (as amended); and
- Provision and Use of Work Equipment Regulations 1998 (PUWER).

HSE employs human factors specialists, including ergonomists and psychologists (occupational, health and cognitive psychologists) to work closely with other relevant specialists enforcing these Regulations.

## Activity 10
## Do you have a problem?

### Objectives

This activity will help you to:

- identify problems in the use of workspace
- evaluate the risk from the working environment and human factors.

### Task

1 The question here is 'Do you have a problem in your work area?'

The easiest way to tackle this is to break it down and examine the key components of problems associated with the human factor. You may be able to list others specific to your workplace and people. Use the following list to help you evaluate the potential risks. An assessment like this can contribute towards your risk assessment of the work environment and human factors.

| *Do you have a problem?* | *Yes or No* | *Number or scale of the problem* | *Potential reasons and hazards* |
|---|---|---|---|
| Have you experienced any major incidents this year? | | | |
| Have accidents been reported? What are they and how serious? | | | |
| Are you aware of any less serious unreported accidents? | | | |
| Are days of through ill-health common? Do you recognise any trends? | | | |
| Do you have a problem with absenteeism? Is it possible to find out what the issues are? | | | |
| Have you noticed any periods of general poor productivity or quality? | | | |
| Does quality tail off at any time during the week? | | | |
| Have you noticed any trends towards lack of concentration? | | | |
| Is PPE being used by staff? Is it being used appropriately? | | | |

2 Your next task is to think about how you monitor the human factors. What processes do you have in place to help you spot trends and patterns, whether seasonal or occasional?

## Feedback

To carry out a review like this it might be helpful to consider the following:

- The job being done and what this means for the employee – what processes and actions do they need to do to fulfil the task?
- The equipment being used – is it fit for purpose or is it contributing to aches, strains or accidents? Is PPE being used effectively
- The physical environment – is it conducive to good working practice and productivity? Are heating, lighting and ventilation adequate? Have you considered noise and vibration levels?
- Whether social factors could make the job or the working environment easier? Are teams supporting each other and are you as a manager approachable about the issues arising?
- What guidance or training have individuals had to support them?

Monitoring in the workplace is about being around to see that processes are being carried out properly and that equipment is being used appropriately. You may not always be able to be out and about with people to assess their working environment in which case it is important that people recognise and respect the culture of safety that you are trying to instill. It is important to make people feel that they are contributing positively if they raise issues with you.

## Legally you must...

...protect the health and safety of everyone in the workplace and ensure that adequate welfare facilities are provided for people at work.

Source: Workplace Health, Safety and Welfare – a short guide for managers, (2004)

## Legal obligations

Employers have a general duty under the HASW Act 1974 to ensure the safety of people at work. However the Workplace (Health, Safety and Welfare) Regulations 1992 (WHSW) Regulations) were brought in to provide more detail and to replace some old laws. The workplace is redefined in these Regulations. Where we used to have Offices, Shops and Railway Premises – we now have a wide-ranging term covering schools, hospitals, hotels or places of entertainment. The term applies to anywhere work is carried out, including out-of-doors and for homeworkers, but does not apply to domestic dwellings.

More details on homeworking can be found in the HSE leaflet INDG 226.

## Health and safety law poster

The HASW Act demands that all places of work employing anyone display a poster. Alternatively, you can provide your employees with individual copies of the same information in a leaflet called *Health and safety law: What you should know*. The poster and leaflets are available from HSE Books.

The Health and Safety Law Poster contains three boxes that need to be completed. In these boxes you need to write the Employee Representative(s) that have been appointed/elected by the employees, or if the employer consults directly with the employees, then this box is left blank. The Management Representative appointed by the employer for health and safety, (i.e. health and safety officer). And finally the appropriate Enforcing Authority, which will depend on the type of business you conduct.

# Insurance requirements

## Compulsory employer's liability insurance

In order to protect employees, an employer is required to have employer's liability insurance. If someone is injured as a result of an accident at work, or becomes ill as a result of work, and if they believe their employer is responsible, they may seek compensation. The insurance is a safeguard to ensure that the employer can adequately compensate the employee. The insurance certificate needs to be displayed on the premises.

Public liability insurance is different. It covers employers for claims made against them by members of the public or other businesses, but not for claims by employees. While public liability insurance is generally voluntary, employer's liability is compulsory.

Uninsured costs can include sick pay, repair and clean up costs, products lost or damaged, cost of replacement staff, extra overtime, fines and damage to corporate reputation. Minimising risks and accidents is essential to reduce what can be considerable insurance premiums and uninsured costs.

# Hygiene and welfare facilities

Hygiene and welfare facilities need to be fit for purpose, promote the comfort of employees and visitors and be appropriately sited on the premises. A review of the facilities might include checking:

- premises and toilets are kept clean, ventilated and in good working order
- where required facilities for washing and drying are provided, including in some areas, drying clothes
- waste bins are emptied regularly
- there is a clean supply of drinking water
- there is space provided for taking food and drinks including a means for washing-up and heating water
- there are facilities to clean up spillages.

Comfort in the workplace contributes to higher productivity and a reduction in stress levels. Regulations are set out in the Workplace (Health, Safety and Welfare) Regulations 1992 that cover:

**Ventilation** – fresh, clean air should be uncontaminated by discharges. Windows may provide sufficient ventilation, but where necessary, mechanical ventilation systems should be provided and regularly maintained.

**Temperatures in indoor workplaces** – comfort depends on air temperature, radiant heat, air movement and humidity. Individual preference makes it difficult to specify a thermal environment, which satisfies everyone. For workplaces where the activity is mainly sedentary, the temperature should normally be at least 16°C. If work involves physical effort, it should be at least 13°C (unless other laws require lower temperatures).

Where staff are required to work in **hot or cold environments** consideration should be given to minimising the risks of heat or cold stress, restriction of exposure time, use of suitable clothing, training, and supervision.

**Lighting** – should be sufficient to allow people to work and move about safely. If necessary, local lighting should be provided at individual workstations, and at places of particular risk, such as crossing points on traffic routes.

Source: Workplace Health, Safety and Welfare – a short guide for managers INDG 244 HSE

## Noise

Noise is a potential problem in the workplace. The Noise at Work Regulations 1989 specify action levels at which the hearing of employees must be protected. Occupational noise can lead to three main acute effects:

- Temporary threshold shift – caused by short excessive noise exposures which result in slight deafness. Deafness is reversible when the noise is removed.
- Tinnitus – a ringing in the ears caused by an intense and prolonged high noise level.
- Acute acoustic trauma – caused by a very loud noise like an explosion. It is usually reversible but severe explosive sounds can cause permanent damage to the eardrum.

Occupational noise can also lead to chronic and permanent versions of the above effects. With sustained exposure to high noise levels individuals can experience permanent hearing loss. Reduction of the noise at source, limiting exposure and personal protective equipment are the principal means by which noise exposure can be reduced.

# To hire or not to hire?

Your facilities and premises are an asset in themselves. Whether they are buildings, fields or a means of transport you will need to consider the implications of hiring them out.

When you decide to hire out premises or facilities to others you are entering a contract which both parties need to understand and agree to. Your responsibility is to hire functional premises, fit for the purpose of hire and suitably risk assessed for health and safety purposes.

The person responsible for hiring the premises should carry out their own risk assessment of the premises relating to the activities they will be used for. Your management responsibility may be to ensure that any risks identified are cross-matched against your own risk assessment and any additional risks to your own employees are minimised.

There may be special conditions that you need to impose on hirers. The following is not an exhaustive list but it will give you an idea of the sorts of issues that can arise.

- Hirers should be required to know, understand and communicate any fire and emergency procedures.
- Hirers will need to know the maximum number of people admitted to the premises and comply with this.
- Hirers are not normally permitted to sub-let the property.
- Arrangements should be made to secure the building after use.
- Any equipment brought on to the site may need to be checked or the hirer may need to be made aware that they are responsible for the safety of the equipment and any insurance liabilities.
- You may wish to make restrictions on materials brought on to the premises and this might include alcohol.

## Activity 11
## Questions about health and safety

### Objectives

This activity will help you to review your understanding of some of the features of health and safety law.

### Task

Answer the following questions:

**Question 1:** What details do you need to put on the new health and safety law poster?

**Question 2:** Are employers required to have public liability insurance?

**Question 3:** What minimum temperature is normally applied to workspaces used for generally sedentary work?

**Question 4:** What are your key responsibilities in hiring out premises?

**Question 5:** What is the main requirement of the Management of Health and Safety at Work Regulations 1999 (the Management Regulations)?

## Feedback

**Question 1:** Employee Representative, Management Representative and Enforcing Authority.

**Question 2:** No, public liability insurance is not a requirement in law, but it is recommended. Compulsory employer's liability insurance is a legal requirement.

**Question 3:** For workplaces where the activity is mainly sedentary, the temperature should normally be at least 16°C.

**Question 4:** Your key responsibilities in hiring out premises are to hire functional premises, fit for the purpose of hire and suitably risk assessed for health and safety purposes.

**Question 5:** The Management of Health and Safety at Work Regulations 1999 (the Management Regulations) generally make more explicit what employers are required to do. The main requirement on employers is to carry out a risk assessment.

# Safe and secure

Safety and security are essential to the safe and effective working of any organisation and are commonly visible to everyone entering a building. Security has two main purposes:

- to protect individuals
- to protect buildings, products, information and equipment.

It is a fact of life that organisations need to provide high levels of security and surveillance.

## Violence and assault

Violence at work causes a lot of stress and in some cases injury. This is not only physical violence but also verbal assault.

The Health and Safety Executive (HSE) defines work-related violence as:

> Any incident in which a person is abused, threatened or assaulted in circumstances relating to their work.
>
> This can include verbal abuse or threats as well as physical attacks.

Source: www.hse.gov.uk/violence/index.htm

The legislation to support and the protect individuals is strong and contained in:

- The Health and Safety at Work etc Act 1974 (HSW Act)
- The Management of Health and Safety at Work Regulations 1999
- The Reporting of Injuries, Diseases and Dangerous Occurrences Regulations 1995 (RIDDOR)
- Safety Representatives and Safety Committees Regulations 1977 (a) and The Health and Safety (Consultation with Employees) Regulations 1996 (b)

Source: www.hse.gov.uk/violence/law.htm

The figures for violence in the workplace are falling, but even at 655,000 they are still far too high. The HSE recommend the following four-point action plan to safeguard staff against the risk of violence:

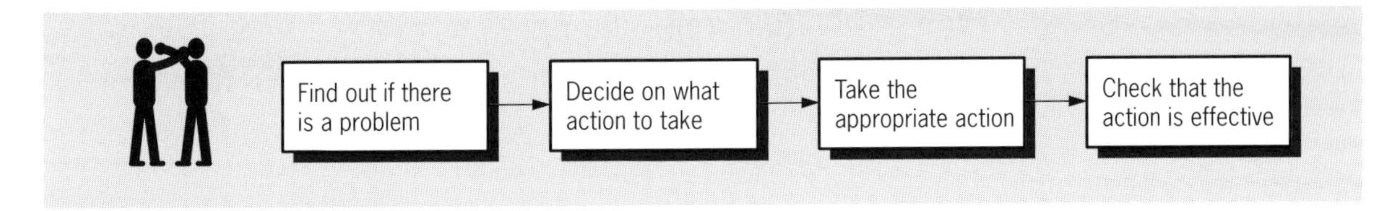


**Figure 3.3** *Safeguarding against violence in the workplace*

Action may include looking at the service provided and the design and layout of the environment. It may mean examining access controls, the use of closed circuit television, alarms, pagers and mobile phones. There are ways of reducing risks by looking at the jobs people do, such as whether you can use cashless payment systems, regular contact with people who work away from their base, avoidance of lone working, transport home after late working.

## Securing the premises

Your role as a manager is to ensure that your team members know and comply with security rules in your organisation. This is about keeping all the resources in your organisation safe, including people. This will include money, materials, products, buildings, vehicles, equipment and information on computers.

In the same way that you protect individuals in the organisation a four-point plan can be applied to the protection of property, information and premises.

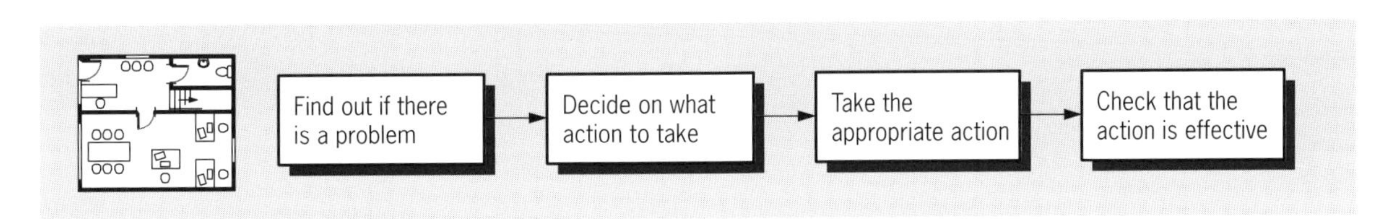


**Figure 3.4** *Protecting the premises*

Again, the actions taken should look at the service provided, the job design and the organisation or layout of the premises. Additionally organisations are increasingly taking action to protect themselves against computer viruses, hackers and electronic attacks. This will involve increased use of password protection, possibly biometric identification and personal electronic signatures. Visual surveillance is also becoming common in organisations.

## Emergency procedures

Being able to respond appropriately in an emergency is an essential consideration in any building design. The procedures used are covered under the Management Regulations 1999 and they aim to limit the damage to people, property and equipment caused by an incident. Local authority fire services will often be prepared to give advice to employers.

Although fire is the most common emergency likely to be faced other possibilities exist, such as:

- explosion from a volatile substance or a bomb
- electrocution
- escape of toxic gases
- dangerous dust like asbestos discharged into the atmosphere
- ram raiding or aircraft crashes
- highly infections diseases like legionnaire's disease
- high winds or flooding.

As a manager, you may be responsible for ensuring that all your employees are familiar with the routes for escape in the event of fire, the procedures in the event of a harmful leak of gases or substances into the atmosphere, or the actions required to prevent disease. The needs in individual premises will vary, but there are a number of basic components when considering the emergency. Employees will need to know:

- the action to be taken on discovering an incident
- how to use the alarm and if appropriate when to call the emergency services
- procedures for stopping machinery in an emergency
- use of fire extinguishers, blankets or masks for instance
- routes for evacuation and assembly points for staff, customers and visitors.

The procedures must take account of any people with special needs. Printed instructions, including staff allocated to specific responsibilities should be displayed throughout the premises. Remember that visitors will also need to be made aware of the processes. It is recommended that the alarm system should be tested on a weekly basis. There is no specific guidance on how often a fire drill is required, but it should be regular and be scheduled to ensure that people on shifts and part-time employees are covered.

## Activity 12

## Security at work

### Objectives

This activity will help you to review your understanding of the risks to safety and security associated with premises for work.

### Task

Read the scenarios that follow and in each situation assess the risks to safety and security.

1 You are reviewing security around the building. Cars are checked on entering via a barrier and full records are kept of entries and exits. However, you have noticed that pedestrians use the car park as a short cut to get to the shops. Around the building are a number of blind alleys.

The risks are:

2 A front reception desk is available for the public to make enquiries. Staff have entry passes in the form of cards which they need to keep with them at all times. Toilets are available in the reception area and a coffee machine. Only one member of staff is likely to be on duty at any one time and all of the offices are on the floor levels above.

The risks are:

## Feedback

The risks that you suggested may have included:

**Scenario 1** Danger to the public in using a car park where vehicles are moving in and out for access to the shops.

Blind alleys where attacks or break-ins may take place.

Checks may be needed on the identities and business purpose of those entering by car.

Staff potentially in danger from public access and more surveillance may be required.

**Scenario 2** Human error involved in losing or dropping security cards.

People working alone are at greater risk.

The toilet facilities mean that people are out of sight by reception staff.

Risk from disgruntled customers.

Risk from public access.

You may have identified some additional risks.

# ◆ Recap

Making the most of your workspace makes financial, social and common sense. This theme looks at the design, ergonomics and security of your space.

**Examine the range and effectiveness of your facilities**

- The influence of your workspace on your employee's work and the image of your organisation on people from outside are easy to underestimate.
- Facilities are examined in relation to the work that you do, the range of people who come into contact with your facilities and the efficiency of the space in terms of planning, access, size and cost.

**Understand the impact of ergonomics, space to work and the human factors that influence behaviour**

- Improvements in the factors which influence behaviour at work can improve health and safety, by reducing the number of accidents and cases of ill-health at work.

- Improvements will also contribute considerable benefits including reducing the costs associated with such incidents and increasing efficiency.

**Review the legal obligations and insurance requirements including location, planning access, size, hiring out and cost factors**

- Your obligation as a manager and employer is to protect the health and safety of everyone in the workplace and ensure that adequate welfare facilities are provided for people at work.
- Here we cover the health and safety law poster, insurance, hygiene and welfare, hiring out premises.

**Develop an understanding of security and safety implications for employees and premises**

- Safety and security are essential to the safe and effective working of any organisation and are commonly visible to everyone entering a building. Security serves to protect individuals, buildings, products, information and equipment.

**Health and Safety Executive (HSE) www.hse.gov.uk**
The HSE provide a wide range of resources covering specific areas of workplace practice and industry specific guidance. These are available online or from HSE Books.

Guidance relevant to this theme includes:

- 'Health and safety regulation – a short guide' hsc 13 (2003)
- 'Workplace health, safety and welfare – a short guide for managers' INDG 244
- 'Homeworking – Guidance for employers and employees on health and safety' INDG 226 (1996 reprinted 2005).
- 'Understanding ergonomics at work – *Reduce accidents and ill-health and increase productivity by fitting the task to the worker'* INDG 90 (2003)
- 'Working with VDUs' INDG 36 (2003)
- 'Getting to grips with manual handling – a short guide' INDG 143 (2004)
- 'Employers' liability (compulsory insurance) Act 1969 – A guide for employees and their representatives' hse 39 (2003)

**HSE 'Essentials of health and safety at work' (1994) 3rd edition, HSE Books**

**HSE 'Health and safety of homeworkers: Good practice case studies', Research Report RR262 (2004) HSE Books**

**Stranks, J. (2005) 7th Edition, *Handbook of Health and Safety Practice*, Prentice Hall**
This guide provides comprehensive coverage of all the new health and safety legislation, including the legal and the practical aspects of health and safety in the workplace and how they can contribute to your wider business objectives.

# 4 Managing equipment

## Functioning at full capacity

Problems with equipment can be very costly. Running at reduced capacity will have an impact on profitability and potentially affect parts of the process further down the line. Managing equipment is about planning, organising people and maintaining equipment effectively.

Equipment functions at full capacity when:

- the right equipment is being used for the task and it is safe to operate
- tasks are planned to be carried out at the right time
- maintenance is planned to occur before a breakdown
- equipment is secured to prevent damage or unauthorised removal
- individuals using the equipment are trained to use it properly.

The consequences of equipment failure can be high in terms of anything from human injury to lost revenue. Minimising the risks and preventing accidents are crucial.

In this theme you will:

- **review your organisational requirements for equipment usage and operating costs**
- **examine methods of capacity planning**
- **identify areas where effectiveness, safety, maintenance and security could be improved**
- **minimise risk from equipment through responsible actions and training.**

## Equipment and capacity

Managing equipment and performing to capacity involves looking for the best use of resources. In this theme we focus on your equipment needs. The starting point is to examine some of the questions to which a manager with responsibility for any equipment and its usage should be able to respond.

| *Equipment usage* | *Yes* | *No* |
| --- | --- | --- |
| Do you look for ways to use time and equipment more cost effectively? | ☐ | ☐ |
| Do you actively review schedules to avoid over-runs or down time? | ☐ | ☐ |
| Do you actively seek ways of improving current systems and procedures for using equipment? | ☐ | ☐ |
| Do you question whether all current equipment is necessary to maintain productivity and quality standards? | ☐ | ☐ |
| Do you question whether you have sufficient equipment to maintain productivity and quality standards? | ☐ | ☐ |
| Will you try a new way to do a job or task because it might be more effective? | ☐ | ☐ |
| Do you actively monitor costs? | ☐ | ☐ |
| Have you established plans to reduce inefficiency? | ☐ | ☐ |
| Do you have systems in place to assure security for equipment? | ☐ | ☐ |
| Do you resist the temptation to supply to a customer's request even though it might be out of the scope of the products and services generally produced or even out of the capacity and competence of those producing it? | ☐ | ☐ |

**Table 4.1** *Examining effective usage of equipment*

If you can answer 'Yes' to all of these questions then you are likely to be actively pursuing efficient and effective use of equipment in your area of responsibility.

The process for managing equipment and maintaining its use at the appropriate level of capacity can be summarised as follows.

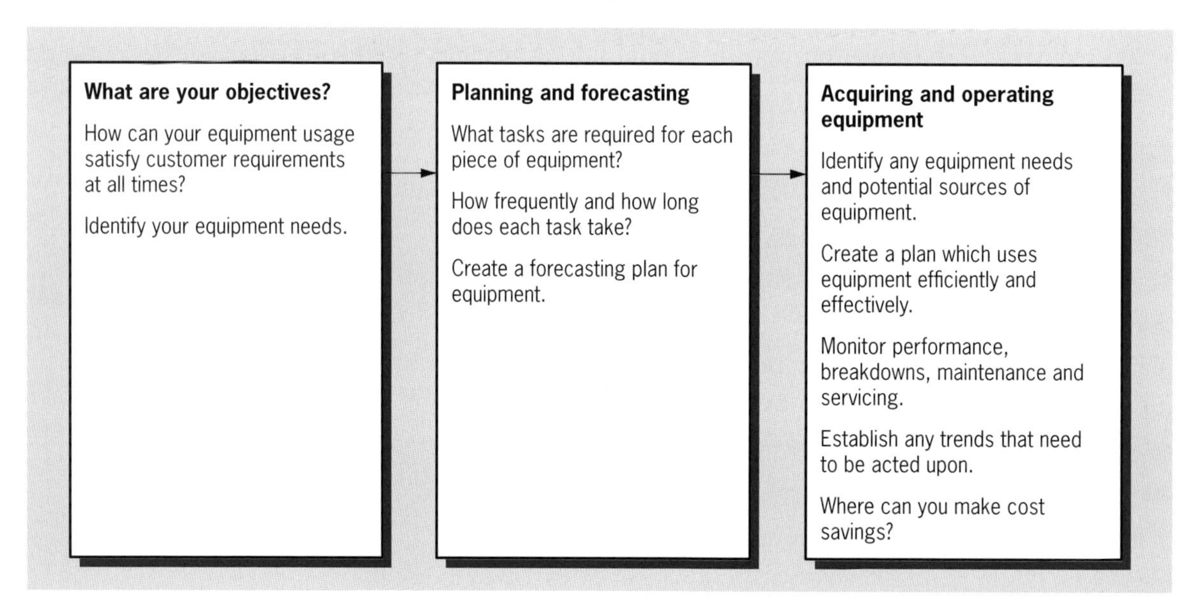


**Figure 4.1** *The process for managing equipment*

The aim is typically to produce the right goods or services, at the right time and at the right cost. It is relatively easy to ensure products are available for the customer if high levels of stock are held or excess capacity is made available. However high levels of stock and capacity are a cost burden to the organisation and may increase prices higher than the customer is willing to pay.

## What are your objectives?

Your objectives form the basis of the tasks and activities that your organisations carry out. There are various constraints on the organisation and you will probably never feel your have the right amount or quality of facilities and equipment that you require. The organisation's ability to purchase resources is constrained by factors such as availability and cost. The resources available to you are constrained by the perceived importance to the organisation as a whole. The aim is to create a situation where you can fulfill the priority objectives for your area of responsibility with effective use of the resources and equipment available to you.

Setting clear objectives for resource and equipment needs is then vital to the presentation of your case for financial resources. Similarly, an understanding of the key resource limitations likely to be imposed upon you is essential for a realistic assessment of your ability to supply goods or services.

## Capacity planning and forecasting

Organisational requirements need to consider:

- the types of work your unit or organisation is engaged in
- the processes or tasks involved in transforming materials into goods or services
- the types of equipment and materials used.

Capacity planning is concerned with bringing materials and equipment together in the transformational process to create the goods that are sold. In this theme we are focusing on equipment and ways that it can be used to efficiently and effectively transform inputs into outputs as well as strategies to manage the flow of materials for Just-In-Time production.

A key element of effective capacity planning is forecasting. All planning has an element of risk because you are dealing with the future. Resource planning is no exception. The organisation plans its resources around what it forecasts to the demand for its products or services, based on projected sales.

A number of techniques exist for forecasting such as time-series analysis, statistical demand analysis, leading indicators, the Delphi Technique and customer intentions surveys. These tools enable you to relate demand to organisational objectives.

### Time-series analysis

Time-series analysis is based on the proposition that past behaviour is a guide to future behaviour; for instance, past sales are a guide to future sales. A time-series is any set of data collected over time – hourly, daily, weekly, monthly, quarterly or annually – and arranged in chronological order. It is then broken down into four categories:

- **Trends** – involving long-term patterns such as future growth.
- **Cycles** – the medium range changes caused by economic conditions and competition.
- **Seasonal patterns** – regular changes during the year such as weather factors or holidays which affect sales.
- **Random occurrences** – one-off events such as closures, regulations and environmental disasters.

### Statistical demand analysis

This technique is based on demand factors other than time-related factors. Typically, a formula is used to predict future demand, as shown in the following example.

For example, the South of Scotland Electricity Board developed an equation that predicted the annual sales of washing machines (Q) to be:

$$Q = 210{,}739 - 703P + 69H + 20Y$$

where $P$ = average installed price; $H$ = new single-family homes connected to utilities; and $Y$ = per capita income.

Thus in a year when an average installed price is £387, there are 5,000 new connected homes and the average per capita income is £4,800, from the equation we would predict the actual sales of washing machines to be 379,678 units:

$$Q = 210{,}739 - 703(387) + 69(5{,}000) + 20(4{,}800)$$

The equation was found to be 95 % accurate.

**Figure 4.2** *Forecasting sales equation* Source: Kotler et al (2002), p 297

### Leading indicators

This is a time series technique involving specific statistics that are associated with your organisation's performance. They indicate future trends and thereby lead you to take action based on the trends. For example, manufacturing output statistics may give you a view of your company's likely demand levels. House prices can indicate to a building company the relative supply and demand for houses.

### Delphi technique

Delphi technique is a qualitative forecasting technique to tap the opinions of experts based on several rounds of written interviews. The feedback is summarised and refined and then the process repeated until a consensus is reached about likely trends.

### Customer intentions' survey

Instead of asking experts the levels of future demand, why not ask the customers? Are they likely to buy such and such a product on a scale of 1-10 within the next three, six, or twelve months?

Forecasting involves making assumptions about demand and these can turn out to be wrong. Some organisations are finding innovative ways to improve the accuracy of their forecasting:

Software manufacturers are collaborating with distributors, retailers and other supply chain partners and swapping point-of-sale information, consumer feedback, and other data over the Internet to improve their forecasts of demand. They are increasingly relying on hard data before committing their product run or, in the case of distributors, before stocking orders. The concept is known as 'pull' manufacturing. This is where, for example, a company may build a PC, but hold back a final version until an up-to-date forecast based on sales data tells it whether customers are more likely to buy a PC with X processing speed or one with Y processing speed.

Source: adapted from *Electronic Buyers News* (2000)

## Changes in demand levels

The relationship of the forecast demand to the equipment and human resources needed to meet the demand is a question of correlation.

Correlation is simply the relationship between two variables.

Based on forecast demand, resource plans indicate what resources are needed and how much they will cost. Because demand usually fluctuates, how can you organise resources and equipment to meet demand? There are broadly three types of organising plan you can use:

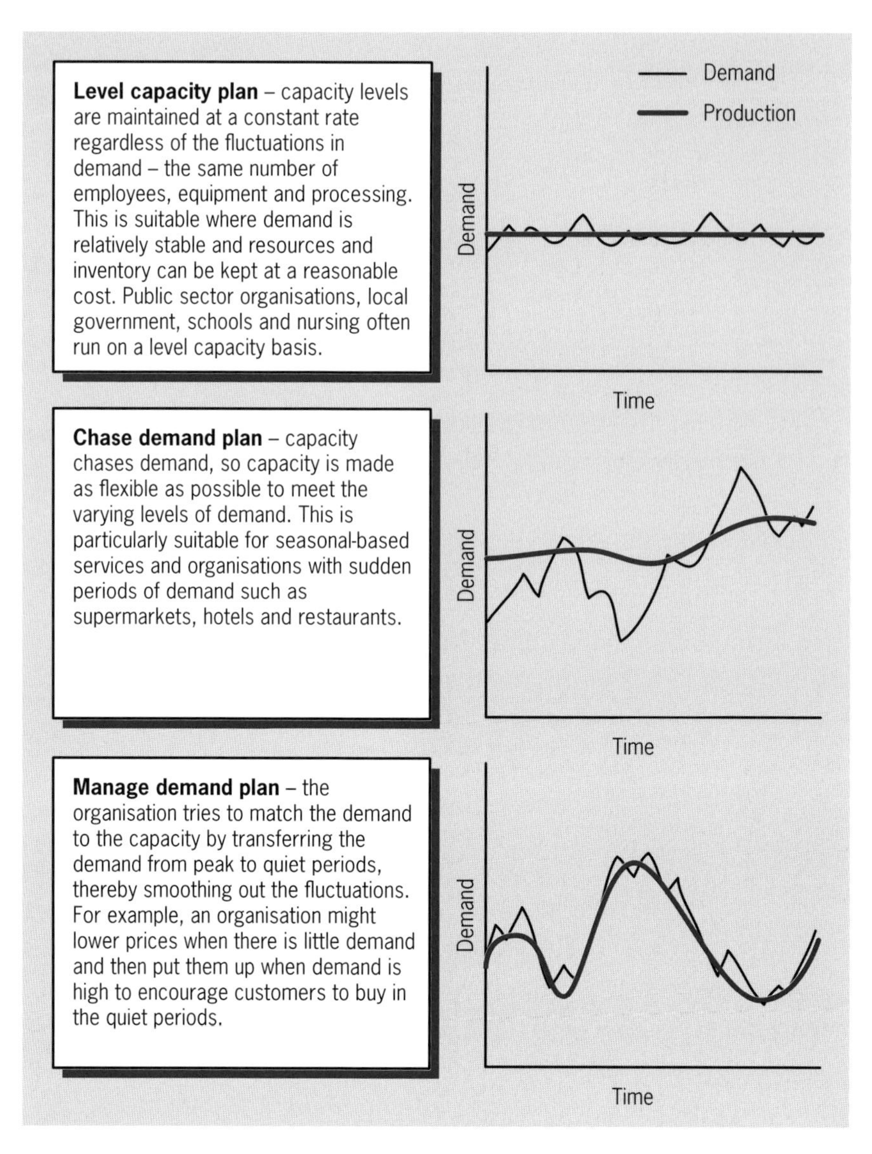


**Figure 4.3** *Planning for resources to meet demand*

Meeting fluctuating demand based on the way you use your resources is called capacity planning. Many organisations actually use a mixture of these plans.

## Acquiring and operating equipment

Mistakes in purchasing equipment and inefficient purchasing systems can hit your organisation's profits.

Acquisition of equipment typically involves a significant outlay and a case may need to be made for large items. The case is normally based on an evaluation of the:

- potential cost of the new equipment
- cost of maintaining any existing equipment
- opportunity cost of replacing or installing new equipment – in other words what savings or increased productivity could be achieved with new equipment

- implications of not replacing or installing new equipment – what would it mean in terms of productivity, safety, operating costs, etc.

## Operating costs

Equipment operating effectively and being well maintained is likely to cost less than the purchase of new equipment. Breakdowns and unplanned stoppages, however, can impact significantly on the ability of the organisation to function appropriately. The performance of equipment needs to be monitored to review operating costs. There are five typical performance criteria that can be applied to review operating costs.

| *Performance criteria* | *Performance indicators for equipment* |
|---|---|
| **Quality** | Number of defective, soiled or damaged units per delivery<br>Orders returned<br>Level of internal or external customer complaints<br>Stock wastage levels<br>Waste levels including energy |
| **Speed** | Order lead time<br>Frequency of delivery<br>Speed of delivery<br>Speed of supply chain – from order to production to distribution |
| **Dependability** | Average lateness of orders or service provision<br>Percentage of orders delivered late<br>Schedule adherence<br>Percentage of products in stock<br>Facilities breakdown levels<br>Product life |
| **Flexibility** | Range of supplies<br>Time to increase supply rate<br>Time to change order<br>Delivery capacity<br>Time to change schedules |
| **Cost** | Stock turn rate (rate of stock turnover)<br>Waste and wastage costs<br>Actual costs v budget projections<br>Operational productivity<br>Discounts |

**Table 4.2** *Performance criteria to review operating costs*

## Lean management

Lean management focuses on controlling and improving processes and incorporates the elimination of waste including time, resources, space, movement etc. Techniques have been inherited from Japanese production companies, which treat people as a key part of the solution to problems of production.

Their approach to scheduling, stock control, flow, maintenance and flexibility was to initiate a problem solving culture. Staff were trained and multiskilled and then the responsibility for operational flow was handed to them. The aim was to find ways to smooth peaks and troughs and maximise the flexibility of the processes.

Efforts were made in the 1980s to identify the source of the significant advantages displayed by this operating system. The major differences lay in the way production was organised and managed rather than more capital investment or equipment that is more modern.

## Activity 13
## Equipment capacity

### Objectives

This activity will help you to:

- find ways to use your equipment more efficiently
- understand the tools that will help you to forecast demand more effectively.

### Task

Answer the following questions and review your equipment capacity.

What are your key work-based productivity or service level targets?

Which, if any, of the following methods do you use to forecast demand or requirements for goods or services?

How effective are your forecasts? Have a look back at the typical performance criteria that can be applied to reviewing operating costs. And how could any of the following techniques improve your forecasting?

| *Demand forecasting methods* | *Effectiveness and potential for improvements* |
|---|---|
| Time-series analysis | |
| Statistical demand analysis | |
| Leading indicators | |
| Delphi technique | |
| Customer intentions surveys | |
| Other techniques | |

How well do, or how could, the following plans to manage changes in demand work in your organisation?

**Level capacity plan**

**Chase demand plan**

Manage demand plan

## Feedback

Your responses will, of course, be specific to your organisation and the types of product or service it offers. The important aspect of this activity is to look at how other approaches could be used and possibly improve effectiveness. It may be worth talking some of these through with your teams to see if they could be implemented.

## Work safe

The safety of equipment is governed by a number of regulations. These particularly apply to the acquisition of equipment and are designed to ensure that equipment meets standards which make it fit for use.

- The Supply of Machinery (Safety) Regulations 1992 SMSR
- Provision and Use of Work Equipment Regulations 1998 PUWER
- Lifting Operations and Lifting Equipment regulations 1998 LOLER
- The Supply of Machinery (Safety) Regulations 1992 as amended by the Supply of Machinery (Safety) (Amendment) Regulations 1994 (SMSR)

These regulations set out the essential requirements, which must be met before machinery or safety components may be supplied in the UK. The indicator or a safety checked piece of equipment under these regulations is a CE marking.

**Provision and Use of Work Equipment Regulations 1998 PUWER**

In general terms, PUWER requires that equipment provided for use at work is:

- suitable for the intended use
- safe for use, maintained in a safe condition and, in certain circumstances, inspected to ensure this remains the case
- used only by people who have received adequate information, instruction and training; and
- accompanied by suitable safety measures, e.g. protective devices, markings, warnings.

**Lifting Operations and Lifting Equipment Regulations 1998 LOLER**

In general, LOLER requires that any lifting equipment used at work for lifting or lowering loads is:

- strong and stable enough for particular use and marked to indicate safe working loads;
- positioned and installed to minimise any risks;
- used safely, i.e. the work is planned, organised and performed by competent people; and
- subject to ongoing thorough examination and, where appropriate, inspection by competent people.

Source: www.hse.gov.uk

Other regulations and guidance that may apply include:

- Personal Protective Equipment at Work Regulations 1992
- Guidelines on Workplace Transport Safety – HSE guidance INDG 199
- Portable electrical equipment – HSE guidance INDG 236
- Using work equipment safely – HSE guidance INDG 229

## Your responsibilities

Anyone with responsibility for equipment has three main responsibilities and they are to ensure that the equipment is in good working order; to ensure that the people using the equipment are appropriately trained to use and check the equipment and report faults and to manage the security of the equipment. Some guidelines for a manager with these responsibilities are to:

- look at all the equipment in use, identify any risks
- evaluate security measures to protect the equipment
- consider what can be done to prevent or reduce these risks

- check whether any of these measures are in place already
- decide whether more needs to be done.

Source: Using work equipment safely – HSE guidance INDG 229

The point of risk assessments and monitoring and health and safety policies is to prevent accidents and ill-health before they happen. Waiting for accidents to happen costs lives, injures people and wastes money.

Developing a work safe culture is an important aspect of an organisation's approach to health and safety and it's an element of safety to which all managers should contribute. Make health and safety an integral part of your management approach. It should form part of your team meetings, discussions, training schedule, assessments, resource allocation and planning, budgeting, performance, contractor relations and equipment purchasing.

## Training for safety

Health and safety considerations are continuous and not a one-off activity. Training and regular updates for safety in the use of equipment are essential in a rapidly changing working environment. Training needs to address four key areas in accident and ill-health prevention.

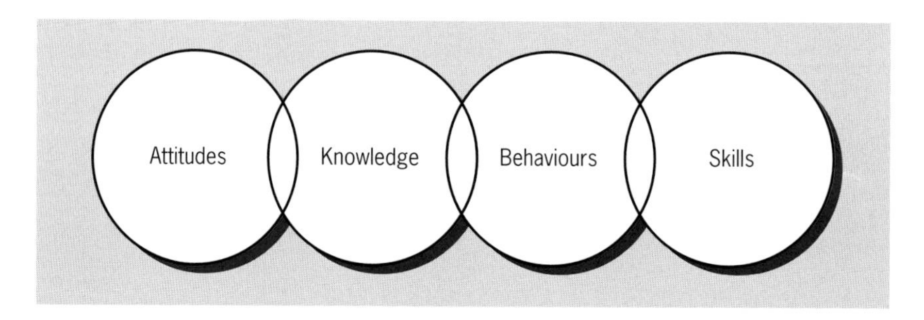


**Figure 4.4** *Elements of training for safety*

### Attitudes

Address attitudes first, employees need to see that this matters to you and to the organisation and why it should matter to them. If you can present them with any statistics relevant to their work area about, absences, ill-health or accidents and how much they cost the organisation this should have a powerful impact.

### Knowledge

Knowledge of the risks and what health and safety encompasses can help to uncover hidden areas of ill-health or potential accidents. For instance examining attitudes and knowledge about stress and absenteeism might help to reveal the particular stresses involved in the workplace and help employees understand that these are recognised conditions and important to the organisation.

## Behaviours

An exploration of the current and desired behaviours and the consequences of carrying on with dangerous practices will highlight the importance of health and safety and illustrate different ways of working which might also be more productive.

## Skills

Having the skills to do the job is an essential requirement. Training should seek to raise issues about skills and the requirements for using equipment or handling materials. Materials or equipment suppliers might be able to help in the delivery of some skills based training. Again, the impact can be felt in improved productivity as well as better health and safety.

## Activity 14
## Safety factors

### Objectives

This activity will help you to:

- assess the risks and potential implications of an obstacle presenting a hazard
- identify areas where effectiveness, safety, maintenance and security could be improved.

### Task

1 You are a manager in a large administration office. You take delivery of a set of new computers for the sales team. Your IT staff start to unpack the computers and put them in place. One of the IT team stumbles into the open drawer of a filing cabinet that is temporarily positioned close to the top of the stair well. He falls downstairs seriously breaking his arm, dislocating his shoulder and trashing the computer.

What are the potential implications of this accident? Think about the personal injury, the costs, the responsibilities and replacement human resources.

What precautions could have prevented this accident?

2 Identify areas where effectiveness, safety, maintenance and security could be improved in your organisation. You may wish to use some of the HSE guidance notes listed above to help you. Try to provide a review of your findings from the following actions.

Look at all the equipment in use, decide what can cause risks, and how.

Evaluate security measures to protect the equipment.

Consider what can be done to prevent or reduce these risks.

Check whether any of these measures are in place already.

Decide whether more needs to be done.

3 Identify what health and safety training needs exist in your area of responsibility and whether they are concerned with attitudes, knowledge, behaviours or skills.

| *Health and safety training needs* | *Attitudes* | *Knowledge* | *Behaviours* | *Skills* |
|---|---|---|---|---|
| | | | | |

## Feedback

1 The potential implications are:

- serious personal and lasting injury to the member of your IT staff
- emotional distress caused to the injured member of staff
- potential loss of income to the member of staff if the recovery is prolonged
- potential compensation claim by the injured member of staff
- higher employer's liability insurance premiums
- the expense of replacing the computer
- the productivity of the business may be affected by the absence of the member of IT staff and loss of a computer
- the cost of training new staff if the member of IT staff is off for a prolonged period
- lost time clearing up and completing the administration associated with the accident
- loss of productivity if other members of staff are involved.

The precautions that could have prevented this accident may have included:

- removing the filing cabinet so that it was not causing an obstruction in a potentially very dangerous position
- checking the route was clear before trying to move the computer equipment
- asking for the advice or help of someone familiar with manual handling guidelines
- assessing the risks associated with the IT staff carrying out this task
- reminding staff of the need for vigilance about dangerously positioned furniture and access routes.

2 Your responses will be individual to your area of responsibility, but consider the range of regulations and guidance available to you. This work can form part of a risk assessment process.

3 Again your responses will be individual to your area of responsibility and individual capabilities. The aim is to promote learning and curiosity by clearly focusing individuals on their responsibilities to themselves, to others and the organisation.

## ◆ Recap

**Review your organisational requirements for equipment usage and operating costs**

- The aim is typically to produce the right goods or services, at the right time and at the right cost.
- You need to manage the situation to fulfill the priority objectives for your area of responsibility with effective use of the resources and equipment available to you.

**Examine methods of capacity planning**

- Capacity planning is concerned with bringing materials and equipment together in the transformational process to create the goods that are sold.
- A key element of effective capacity planning is forecasting. All planning has an element of risk because you are dealing with the future.
- A number of techniques exist for forecasting such as time-series analysis, statistical demand analysis, leading indicators, the Delphi Technique and customer intentions surveys. These tools enable you to relate demand to organisational objectives.

**Identify areas where effectiveness, safety, maintenance and security could be improved**

- Anyone with responsibility for equipment has three main responsibilities and they are to ensure that the equipment is in good working order; to ensure that the people using the equipment are appropriately trained to use and check the equipment and report faults and to manage the security of the equipment.

**Minimise risk from equipment through responsible actions and training**

- Health and safety considerations are continuous and not a one-off activity. Training and regular updates for safety in the use of equipment are essential in a rapidly changing working environment.

**Health and Safety Executive (HSE) www.hse.gov.uk**
The HSE provide a wide range of resources covering specific areas of workplace practice and industry specific guidance. These are available online or from HSE Books.

Guidance relevant to this theme includes:

- 'Health and safety regulation – a short guide' HSC 13 (2003)
- 'Workplace health, safety and welfare – a short guide for managers' INDG 244
- 'Working with VDUs' INDG 36 (2003)
- 'Getting to grips with manual handling – a short guide' INDG 143 (2004)
- 'A short guide to the Personal Protective Equipment at Work Regulations 1992' INDG 174 (2005)
- 'Maintaining portable electrical equipment in offices and other low-risk environments' INDG 236 (1996 reprinted 2004)
- 'Using work equipment safely – guidance' INDG 229 (2002 reprinted 2003)

**HSE (2001) *Reducing Risks, Protecting People – HSE's decision-making process*, R2p2 HSE Books**

**Hughes, P. and Ferrett, E. (2005) 2nd Edition, *Introduction to health and safety at work*, Elsevier Butterworth-Heinemann**
The Introduction to health and safety at work provides a clear outline of all occupational safety and health. It is particularly useful

to managers, focusing on their responsibilities in the workplace. It covers the essential elements of health and safety management, the legal framework and risk assessment. Particularly relevant to this theme, there are sections on the principles of control, work equipment and electrical hazards and control and monitoring and reviewing health and safety.

**Brown, S., Lamming, R., Bessant, J. and Jones, P. (2000) *Strategic Operations Management*, Elsevier Butterworth-Heinemann**
This text works from the premise that successful operations management depends on developing strategies which combine manufacturing and service expertise and competencies to provide an effective service to customers. It covers areas such as supply management, capacity planning and materials management as well as quality.

# 5 Managing materials

## Making materials work for you

Anything is possible if you have all the materials and people to do the job available all the time. However there is a cost involved that is likely to prohibit a simple process of over purchasing materials, just in case they are required. The skill of management lies in having the right materials in the right place at the right time for people to use.

This theme identifies ways of determining materials requirements and purchasing procedures that support processes such as Materials Resource Planning (MRP) and Just In Time (JIT) production. Having got the materials, we then consider receipt, storage and issue of materials and some of the legislation associated with safe storage and distribution.

Minimising waste is essential in an environment of over consumption. Most organisations are becoming increasingly aware of their own environmental impact and customers concern for conservation of scarce resources. Here we look at measures that can be taken to minimise waste.

In this theme you will:

- **examine processes to determine materials requirements**
- **identify the impact of storing and securing materials and the legislation associated with storage**
- **review process for receipt and issue of materials including control processes**
- **develop a plan to minimise waste within your area of responsibility.**

## The right materials

Most organisations face a common problem. Customers want goods or services to be available quicker than you can produce them. This necessitates some planning of goods and stocks held. Some of the main considerations in materials planning are as follows:

- How long does it normally take for stock to be delivered once ordered (the lead time)?
- What is the cost and expected pattern of use? It may be acceptable to hold large stocks of low value materials, but for high value items you will need to know the demand in advance.

- On what basis do you predict or forecast your demand? Some examples of forecasting techniques are time-series analysis, statistical demand analysis, the Delphi technique etc.

In this theme we explore two main processes for materials planning, materials requirements planning (MRP) and Just-In-Time management.

## MRP

MRP is a software based production planning and inventory control system used to manage manufacturing processes. It involves the calculation of the purchasing needs of the organisation based on its current stock levels and the predicted future requirements. An MRP system is intended to meet three objectives:

- Ensure materials and products are available for production and delivery to customers.
- Maintain the lowest possible level of inventory.
- Plan manufacturing activities, delivery schedules and purchasing activities.

The questions it provides answers for are: WHAT items are required, HOW MANY are required and WHEN are they required by.

To be effective the MRP system needs a range of inputs and outputs.

### Inputs

- An overall production schedule. This is a combination of all the known and expected demand for the products being created. The schedule spans a period that includes the present and extends several months, and sometimes even years, into the future. It includes how much is required at a time, of what end products or services and the quantities to meet demand.
- Inventory status records. Records of materials available for use already in stock (on hand) and materials on order from suppliers.
- Bills of materials. Details of the materials, components and subassemblies required to make each product.
- Planning Data. This includes all the resources and directions to produce the end items.

### Outputs

The outputs from the MRP are Recommended Production Schedules and Recommended Purchasing Schedules, providing a clear structure for manufacturing and for purchasing. The MRP can also be set up to generate messages and reports, purchase orders, reschedule notices to cancel, increase, delay or speed up existing orders. The recommended schedules should of course be monitored by trained people to group orders to achieve freight savings and to add flexibility.

## Just-in-time

Just-in-time (JIT) resource management aims to reduce stock levels to those needed for immediate use and thereby reduce the costs associated with storage. This is achieved by working closely with suppliers, so that goods arrive just before they are needed. Relationships with suppliers are necessarily very important. Good relationships mean reliable delivery, meeting expected quality standards and good channels for communication. JIT brings with it a need for a complete understanding of the supply needs of an organisation.

The technique was first used by the Ford Motor Company as described explicitly by Henry Ford's My Life and Work (1922):

> 'We have found in buying materials that it is not worth while to buy for other than immediate needs. We buy only enough to fit into the plan of production, taking into consideration the state of transportation at the time. ... That would save a great deal of money, for it would give a very rapid turnover and thus decrease the amount of money tied up in materials.'

Henry Ford

### Toyota's experience

The chief engineer at Toyota's car manufacturing base in Japan in the 1950s, was Taiichi Ohno. He examined accounting assumptions and realised that the factory could be made more flexible, reducing the overhead costs of retooling and reducing the economic lot size to the available warehouse space.

Some of the results were unexpected. A huge amount of cash appeared, apparently from nowhere, as in-process inventory was built out and sold. This by itself generated tremendous enthusiasm in upper management.

Another surprising effect was that the response time of the factory fell to about a day. This improved customer satisfaction by providing vehicles usually within a day or two of the minimum economic shipping delay.

Also, many vehicles began to be built to order, completely eliminating any risk that they would not be sold. This dramatically improved the company's return on equity by eliminating a major source of risk.

Source: http://en.wikipedia.org/wiki/Just_in_time viewed March 2006

## Just in Time strategies

Just in Time is more than just a manufacturing system and the JIT philosophy has also been applied to other segments and other types of manufacturing and service environments. In the commercial sector, it has meant eliminating warehouses in the link between a factory and a retail establishment or using the power of the internet to reach customers directly.

To try and sum it up, it is a fluid combination of logical thinking and the collaboration of people and ideas for improvement. There is a range of strategies that can help in its implementation.

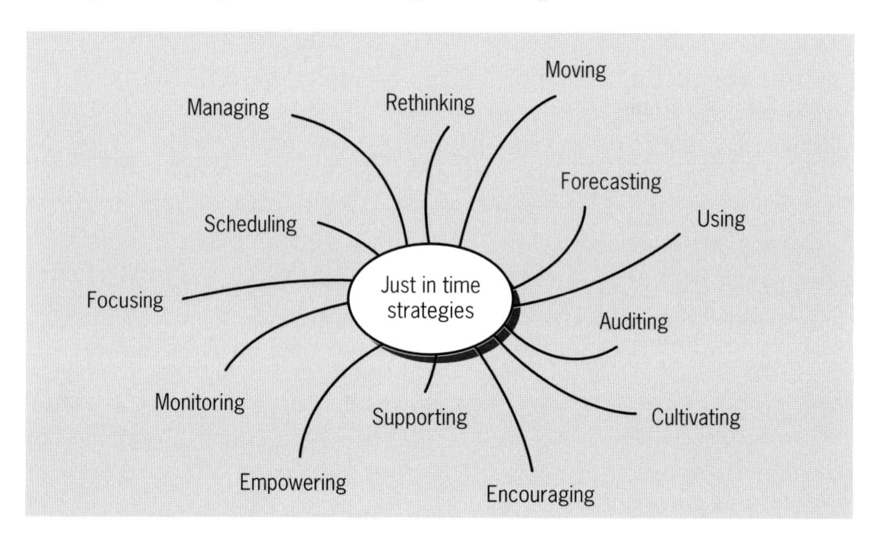


**Figure 5.1** *JIT strategies*

| | |
|---|---|
| **Managing** | To avoid haphazard implementation, someone needs to keep an eye on the system-wide use of equipment and materials, noting breakdowns, lurches, delays and problems. |
| **Rethinking** | Based on experience, some strategies will prove more successful than others and the process must be reviewed by those who know how it works to find alternative methods of working. This will lead to modification, and invention of new approaches and combinations of strategies. |
| **Moving** | Put the equipment, materials and resources where they will do the most good and move them as needs shift. |
| **Grouping** | Group and categorise tasks and components. By grouping you can achieve time efficiencies and minimise wastage. |
| **Forecasting** | Forecasting needs to be accurate to anticipate needs likely to arise and therefore schedule to meet demand. |
| **Scheduling** | Who gets what when? It is a question of deciding and communicating who will need how much equipment at which times. |
| **Focusing** | Focus on the priority areas. Focus on areas where savings can be made in time and resources. |
| **Using** | Make sure no equipment or people are sitting idle. |
| **Auditing** | Gather data on usage of resources and productivity, the amount of waste generated and the amount of stocks held. |
| **Monitoring** | Keep track of how much of your resources are being used, the trends for usage and the work patterns. |
| **Empowering** | Let the people who know make the decisions. As a manager your responsibility is to guide and to recognise value in the contributions, ideas and suggestions of others in your team. |
| **Encouraging** | Encourage those who use resources and provide services to suggest improvements. |
| **Supporting** | Support people to implement changes. |

**Table 5.1** *JIT strategies*

## Activity 15
## Your JIT improvements

### Objectives

This activity will help you to:

- apply JIT techniques and planning to your area of responsibility
- identify materials requirements to meet your organisational needs.

### Task

Using the strategies associated with JIT systems, review the benefits that could help to improve productivity, quality and planning in your situation. Remember that JIT is more of a philosophy these days than purely a manufacturing tool. Use the following headings to direct your thoughts.

| *JIT strategies* | *Steps to improvement* |
|---|---|
| Managing | |
| Rethinking | |
| Revising | |
| Moving | |
| Grouping | |
| Forecasting | |
| Scheduling | |
| Focusing | |
| Using | |
| Auditing | |
| Monitoring | |

| *JIT strategies* | *Steps to improvement* |
|---|---|
| Cultivating | |
| Empowering | |
| Encouraging | |
| Supporting | |

Think about how this impacts on any materials requirements that you may have. Could your requirements be more flexible or be managed differently as a result of your review?

## Feedback

This is a long list of strategies and they may not all be appropriate to your situation. It is however, a good indication of the range of methods involved in a dynamic review of materials and resource planning. It will also help you to clearly identify your materials requirements, any changes and any trends if carried out in consultation with others.

# In the right place

## Purchasing and procurement

Purchasing and procurement is about getting the right materials to the right place at the right time and at the right cost. This involves a process of ensuring that materials can be purchased from suppliers of the right:

- quality
- quantity
- price and payment terms

- range
- flexibility
- compatibility.

Quality refers to product functionality such as durability, reliability, precision, ease of operation and repair, energy efficiency, low maintenance or safety. Compatibility means the way a product or service works with existing production processes. Flexibility refers to the different purchasing options such as online purchase.

### Suppliers

You may need to consider whether to use the same supplier for different resources or even different suppliers for the same resources. An advantage of single sourcing is that you get to know your supplier and can form a stable, enduring relationship. You may also get volume discounts. A drawback is that you may be left stranded if supplies suddenly dry up. Using multiple suppliers eliminates the risks of a cessation of supply and you may be able to drive prices down by competitive tendering. On the other hand, you may find it more difficult to co-ordinate many sources of supply.

You could also strike a balance between one supplier and many by having a small list of preferred suppliers.

### Specifying your requirements

You will want to be sure that any supplier is aware of your requirements. The best way to do this is to write a specification or an invitation to tender (ITT) where appropriate. The ITT asks your supplier to tend a quotation and proposal to show it can meet your criteria and requirements.

### Service level agreement (SLA)

An SLA is an agreement between a supplier or service provider and a customer that specifies the quality of service expected by the customer and promised by the supplier. It will define the service to be provided by the supplier, the responsibilities of the purchaser – for instance for ordering or paying invoices – performance criteria, how performance will be measured and how communications, including complaints, will be managed. There may be a section that illustrates the constraints, i.e. circumstances which may prevent delivery to standards and how to avoid them.

SLA's should not be vague, or too detailed. They shouldn't ignore the cost of monitoring the service levels and they should be reviewed regularly.

## Purchasing

When you have your suppliers you need to put in place a purchasing process to make the exchange of products efficient. The stages in the process are illustrated below.

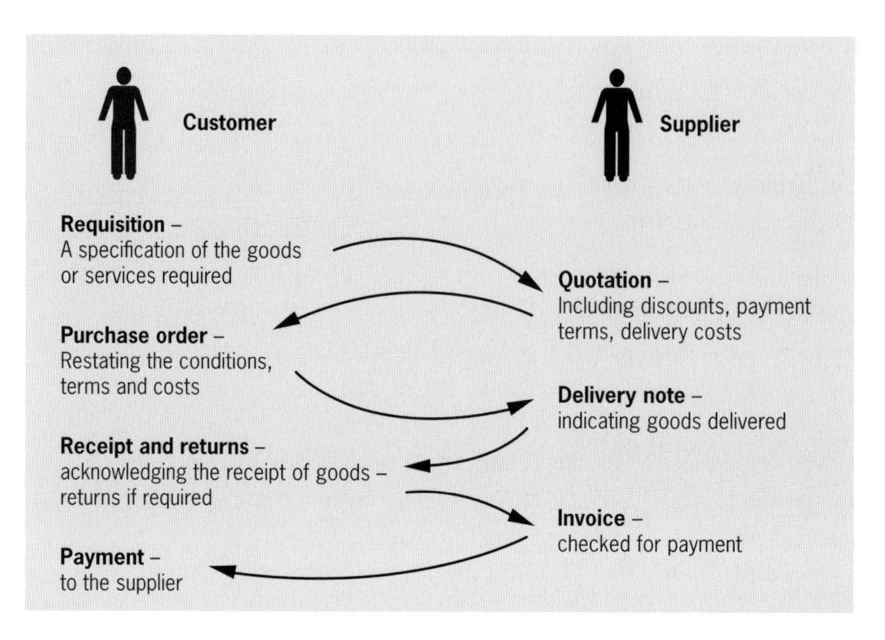


**Figure 5.2** *Stages in the purchasing process*

There is a range of software packages available to support purchasing and you may be required to use these. Whichever one your organisation uses they are only ever as good as the specification you make for the goods.

## Materials control principles

The amount of goods or services that you buy at anyone time will have an impact on the cost of materials. You can control costs more effectively if you have an understanding of EBQ (Economic Batch Quantity), also called Optimal Batch Quantity. This is a supply side measure used to determine the quantity of units that can be produced in any given scenario. So for instance ordering three chairs may entail a wait, since the chairs are made in batches and this is an uneconomic batch. Ordering 30 chairs constitutes a batch and the order can be manufactured immediately.

Conversely, on the purchasing side Economic Order Quantity (the EOQ Model) is a model that defines the optimal quantity to order that minimises total variable costs required to order and hold inventory.

The model was originally developed by F. W. Harris in 1915, though R. H. Wilson is credited for his early in-depth analysis of the model. It roughly equates to:

Total cost = purchase cost + order cost + holding cost

There is an explicit recognition here of the cost of storage for materials, the aim being to minimise the damage that could occur through running out of stock against viable ordering and storage quantities.

Stock costs include:

**Storage costs** Storage entails lighting, heating or cooling, energy, equipment, space and labour.

**Obsolescence costs** Most stored goods have a shelf life – they may perish or go out of date.

**Working capital costs** These arise because of the lag between buying supplies and selling goods. The longer the gap, the more you are tying up capital.

**Organisational costs** The more efficient the stock control system the less likely that goods will be lost, damaged or pilfered.

## Stock control

Efficient stock control can improve your revenue and reduce your costs.

A stock control system is normally designed to facilitate efficient movement of goods in and out. It serves to classify all of the stock and arrange it in areas that make sense for later dispatch. The control system will often prioritise stock by the frequency of its usage and its value. It should be able to account for exactly what you have in stock at any time, identifying any stocks received or expected and materials in production or sold.

## Records

In managing your resources you are likely to use a wide range of documents including contracts, purchase orders, invoices, bills, delivery notes, stock records, SLAs, transfer notes and maintenance records. Record keeping is an important part of any monitoring and control process. These records will help you to:

- comply with legislation, such as health and safety and environmental legislation
- provide evidence of the efficiency and accuracy of the operations
- prove that goods have been received or delivered and checks and actions have taken place
- prove what actions and specifications have been agreed
- prove ownership
- enable efficient ordering, auditing and accounting.

# Activity 16
## Supplier selection

### Objectives

This activity will help you to:

- ensure the availability of the right materials for operations
- review suppliers against key criteria.

### Task

1 Assess three of your current suppliers on a scale of 1 – 5 (where 1 is poor and 5 is excellent) against each of the criteria in the table below.

| *Supplier* | *Quality* | *Quantity* | *Price and payment terms* | *Range* | *Flexibility* | *Compatibility* |
|---|---|---|---|---|---|---|
| 1 | | | | | | |
| 2 | | | | | | |
| 3 | | | | | | |

2 Are your suppliers able to ensure, as far as they are able, that you get the right materials in the right place?

### Feedback

You might consider how satisfied your organisation as a whole is with these and other suppliers. You may need to discuss this with other managers and see if the criteria can be applied more widely.

# Safety and energy management

## Safety and security

The safety and security of materials involves more than protection from loss, theft or pilfering, although these can be important considerations. There are a number of Regulations associated with the storage and security of materials.

### CoSHH

CoSHH is the Control of Substances Hazardous to Health and relates to the Regulations of 2002. Hazardous substances include anything from bathroom bleach to nuclear fuel. The law requires employers to control exposure to hazardous substances to prevent ill-health.

Full details on the regulations are available from the HSE and a guide is provided entitled CoSHH: A brief guide to the Regulations INDG 136.

The basic steps you are required to take are as follows:

Step 1 Assess the risks

Step 2 Decide what precautions are needed

Step 3 Prevent or adequately control exposure

Step 4 Ensure that control measures are used and maintained

Step 5 Monitor the exposure

Step 6 Carry out appropriate health surveillance

Step 7 Prepare plans and procedures to deal with accidents, incidents and emergencies

Step 8 Ensure employees are properly informed, trained and supervised.

## COMAH

The Control of Major Accident Hazards Regulations 1999 (COMAH) and their ammendments 2005, are applicable to any organisation storing or otherwise handling large quantities of industrial chemicals of a hazardous nature. Types of establishments include chemical warehousing, chemical production facilities and some distributors.

The principle aim of the regulations is to reduce the risks of potential major accidents. The regulations operate on two levels' Lower Tier' and 'Upper Tier', determined by the nature of the stock held.

Lower tier establishments are required to document a Major Accident Prevention Policy which should be signed off by the managing director. An upper tier COMAH establishment is required to produce a full safety report which demonstrates that all necessary measures have been taken to minimise risks posed by the site with regard to the environment and local populations. The Regulations are enforced by The HSE, the Environment Agency and SEPA in Scotland.

## CHIP

CHIP refers to the Chemicals (Hazard Information and Packaging for Supply) Regulations 2002. These are sometimes also known as CHIP3. CHIP is the law that applies to suppliers of dangerous chemicals. Its purpose is to protect people and the environment from the effects of those chemicals by requiring suppliers to provide information about the dangers and to package them safely.

CHIP requires the supplier of a dangerous chemical to:

- identify the hazards (dangers) of the chemical. This is known as 'classification'
- give information about the hazards to their customers. Suppliers usually provide this information on the package itself (e.g. a label) and, if supplied for use at work, in a safety data sheet (SDS)
- package the chemical safely.

Source: HSE www.hse.gov.uk/chip/

### NONS

The Notification of New Substance Regulations 1993 (NONS 93) are a set of regulations that seek to protect people and the environment from the possible harmful effects of new substances and to create a 'single market' in new substances across the European Community. NONS 93 aims to identify the possible risks posed to people and the environment from placing new substances on the market. It does this by obtaining information about them in a systematic way so that users may be made aware of the dangers and if necessary, recommendations for control can be made.

Source: HSE www.hse.gov.uk/nons/nons2.htm

These regulations may or may not apply directly to the activities of your organisation. However, you may need to consider your proximity to other organisations where these Regulations are of direct relevance, the types of potentially toxic agents you have on the premises and any requirements for packaging and storage of goods.

## Energy management

The resources used by your organisation, for which you also have responsibility, include the energy and utilities that you use. Energy efficiency and waste control both protect the environment and can make a significant difference to an organisation's profitability.

Here's an example

Increasing the rate of ventilation may increase productivity according to research by Professor Donald Milton of the Harvard School of Public Health. The professor found that when the amount of ventilation provided was increased, the rate of short-term sick leave was lower. He worked out the savings to the organisation from supplying more ventilation to cut down sick leave was $400 per employee per year!

Source: *Energy User News*, 2002

Energy management concerns the control of heating, lighting, ventilation, refrigeration – anything which uses oil, gas electricity, etc. to power it. An energy management system should include:

- making energy inspections to look for waste – such as lights left on unnecessarily, thermostats set too high, toilet facilities which 'over-flush'
- installing more efficient energy systems

- negotiating contracts with suppliers to ensure best possible tariffs and prices
- informing employees of ways to reduce energy use and encouraging them to find ways of reducing energy and waste costs
- an energy management information system to log inspections, validate energy bills, and provide energy reports.

As well as reducing the level of waste energy, you can also reduce the level of physical waste you produce.

A useful rule of thumb for waste reduction is the following which is known as the 'waste hierarchy', i.e. 1 is best, but if you can't do 1, then 2 is next best, and so on.

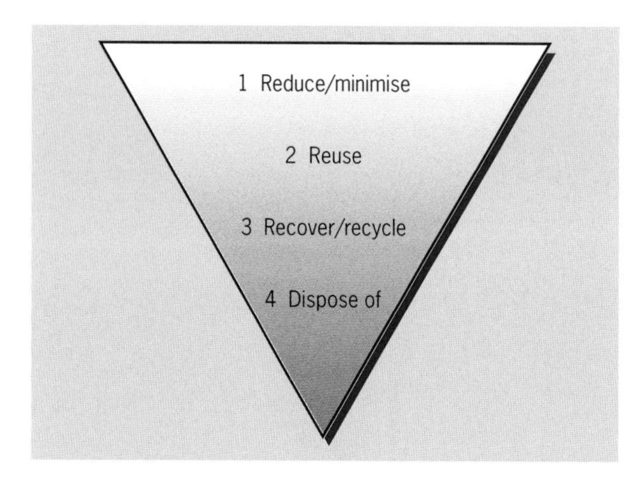


**Figure 5.3** *The waste hierarchy*

The best option is to reduce the waste at source, e.g. by using less packaging with delivered supplies. Next best is to reuse, for example by sending back old toner cartridges for refilling. Next best is to recycle as when you use bottle banks or composters. Disposal to landfill or incineration is considered a final resort where all other options have 'failed'.

## Legal requirements

It's worth remembering that you have a waste management duty of care under the Environmental Protection Act 1990. This means that you have to:

- prevent others illegally handling your waste products
- contain any waste destined for disposal, i.e. store it safely
- ensure waste products are transferred to authorised people only, so that it is disposed of properly
- describe waste destined for transfer accurately – you should be using documents known as 'transfer notes'. Or, if you have a type of dangerous waste known as 'special waste', you should be using 'consignment notes'. Special waste includes dangerous chemicals and hospital clinical waste.

# Activity 17
## The energy management audit

### Objectives

This activity will help you to:

- audit your current energy usage
- identify potential areas for improvement.

### Task

1 Carry out an energy management audit on your organisation or work area using the table below.

| *Energy management audit* | *Yes* | *No* |
|---|---|---|
| **Transport** | | |
| Are vehicles properly serviced, maintained and tuned? | ☐ | ☐ |
| Do employees share vehicles when they are travelling to the same place on business? | ☐ | ☐ |
| Do some drivers appear to use too much fuel? | ☐ | ☐ |
| Do some drivers need instruction in fuel economy? | ☐ | ☐ |
| Is the most cost-effective form of transport used? | ☐ | ☐ |
| **Lighting** | | |
| Are the most efficient light bulbs being used? | ☐ | ☐ |
| Could more use be made of daylight by moving workstations nearer windows? | ☐ | ☐ |
| Are lights switched off when rooms are not in use? | ☐ | ☐ |
| **Heating** | ☐ | ☐ |
| Is the heating system serviced regularly? | ☐ | ☐ |
| Are thermostats functioning correctly and are they set to the correct temperature? | ☐ | ☐ |
| Is the heating switched off or turned down when the building is empty? | ☐ | ☐ |
| **Air conditioning** | | |
| Is there really a need for it? | ☐ | ☐ |
| Is the system kept clean and regularly maintained? | ☐ | ☐ |
| Is it working against the heating system? | ☐ | ☐ |
| **Insulation** | | |
| Are the wall and roof insulation materials of the correct type and thickness? | ☐ | ☐ |
| **Ventilation** | | |
| Do employees open doors or windows to cool the place down rather than turning down thermostats? | ☐ | ☐ |
| Are there excessive draughts from badly fitting doors and windows? | ☐ | ☐ |
| **Equipment/machinery** | | |
| Is machinery running efficiently? | ☐ | ☐ |
| Could any heat/energy produced by processes be re-used? | ☐ | ☐ |
| Is the right size of machine used for each job? | ☐ | ☐ |
| Are computers turned off when not in use? | ☐ | ☐ |

Source: 'Setting up an Energy Management Scheme', *Operations & Quality Management*, 1999, pp 95-6

2 Develop three actions to minimise waste in your area of operations based on your energy audit findings.

1

2

3

## Feedback

This activity will help you to identify where you need to make improvements. Consider your energy and environmental priorities. What improvements can you make? Discuss your ideas with colleagues in the organisation. Even small actions can have big effects.

This is also a key area for any organisation to gain competitive advantage. Your role as a manager is to explore the options and the costs and time associated with them.

# ◆ Recap

**Examine processes to determine materials requirements**

- In this theme we explore two main processes for materials planning, materials requirements planning (MRP) and Just-In-Time management.

**Identify the impact of procuring, storing and securing materials**

- Purchasing and procurement involves ensuring that materials can be purchased from suppliers of the right quality, quantity, price, range, flexibility and compatibility.
- You can control costs more effectively if you have an understanding of EBQ (Economic Batch Quantity) and Economic Order Quantity (the EOQ Model).
- Efficient stock control can improve your revenue and reduce your costs.

**Develop a plan to minimise waste within your area of responsibility**

- Energy efficiency and waste control both protect the environment and can make a significant difference to an organisation's profitability.

## More @

**Health and Safety Executive (HSE) www.hse.gov.uk**
The HSE provide a wide range of resources covering specific areas of workplace practice and industry specific guidance. These are available online or from HSE Books.

Guidance relevant to this theme includes:

- 'COSHH: A brief guide to the Regulations' INDG 136 (2002)
- 'Major accident prevention policies for lower-tier COMAH establishments' (1999)
- 'The idiots guide to CHIP 3' INDG 350 (2003)

**Brown, S., Lamming, R., Bessant, J. and Jones, P. (2000) *Strategic Operations Management*, Elsevier Butterworth-Heinemann**
This text works from the premise that successful operations management depends on developing strategies which combine manufacturing and service expertise and competencies to provide an effective service to customers. It covers areas such as supply management, capacity planning and materials management as well as quality.

# References


Brown, S., Lamming, R., Bessant, J. and Jones, P. (2000) *Strategic Operations Management*, Elsevier Butterworth-Heinemann

CMI (1999) *Operations & Quality Management*, Chartered Institute of Management

*Electronic Buyers News* (2000) 'Filling in the Forecasting Piece of the SCM Puzzle – A variety of software solutions, some web-based, are available', CMP Publications Inc, vol 61, May

*Energy User News*, (2002) Business News Publishing Co, July 2002, vol. 27, issue 7

Ford, H. T., Nevins and Hill, (1922) *My Life and Work, The Times, the Man, the Company*, TMC

HSE, (2000) 2nd Edition, *Successful health and safety management*, HSG 65, HSE Books PO Box 1999, Sudbury, Suffolk CO10 2WA.

HSE (1994) 3rd edition, *Essentials of health and safety at work*' HSE Books

HSE (2004) *Health and safety of homeworkers: Good practice case studies*, Research Report RR262, HSE Books

HSE, (2001) *Reducing Risks, Protecting People – HSE's decision-making process*, R2p2, HSE Books

HSE, *Investigating accidents and incidents – a workbook for employers, unions, safety representatives and safety professionals* HSG245 HSE Books.

Hughes, P. and Ferrett, E, (2005) 2nd Edition, *Introduction to health and safety at work*, Elsevier Butterworth-Heinemann

Kotler, P., Armstrong, G., Saunders, J. and Wong, V. (2002) 3rd edition, *Principles of Marketing*, FT Prentice Hall

Ridley, J. (2004) 3rd Edition, *Health and safety in brief*, Elsevier Butterworth-Heinemann

Royal Society for the Prevention of Accidents www.rospa.com

Stranks, J. (2005) 7th Edition, 'Handbook of Health and Safety Practice', Prentice Hall


**Guidance leaflets from HSE**


'Health and safety regulation – a short guide' HSC 13 (2003)

'Workplace health, safety and welfare – a short guide for managers' INDG 244

'Directors' responsibilities for Health and Safety INDG343 (2001)

'Health and safety law – what you should know' (1999) This is a short guide for employers and employees.

'A guide to measuring health and safety performance' (2001)

'Consulting Employees on Health and Safety: a Guide to the Law' INDG 232 (2002)

'Five steps to risk assessment' INDG 163 (2003)

*'RIDDOR explained'* HSE 31 (1999, reprinted 2004)

'Homeworking – Guidance for employers and employees on health and safety' INDG 226 (1996 reprinted 2005).

'Understanding ergonomics at work – *Reduce accidents and ill-health and increase productivity by fitting the task to the worker'* INDG 90 (2003)

'Working with VDUs' INDG 36 (2003)

'Getting to grips with manual handling – a short guide' INDG 143 (2004)

'Employers' liability (compulsory insurance) Act 1969 – A guide for employees and their representatives' HSE 39 (2003)

'A short guide to the Personal Protective Equipment at Work Regulations 1992' INDG 174 (2005)

'Maintaining portable electrical equipment in offices and other low-risk environments' INDG 236 (1996 reprinted 2004)

'Using work equipment safely – guidance' INDG 229 (2002 reprinted 2003)

'COSHH: A brief guide to the Regulations' INDG 136 (2002)

'Major accident prevention policies for lower-tier COMAH establishments' (1999)

'The idiots guide to CHIP 3' INDG 350 (2003)

WITHDRAWN
FROM STOCK